The Technological Muse

The Technological Muse

Katonah Museum of Art

November 11, 1990 – February 3, 1991

Curated by Susan Fillin-Yeh

Katonah Museum of Art
Route 22 at Jay Street
Katonah, New York 10536

ISBN:915171-19-8

Library of Congress Catalogue Card Number:
90-061975

Exhibition Coordinators: Angela Gilchrist, Nancy Wallach
Docent Trainer: Nancy Wallach
Education Consultant: Yvonne Pollack
Public Programs: Myrna Clyman, Betty Horowitz
Exhibition Design: Arthur M. Clark
Catalogue Design: Marc Zaref Design, New York
Publicity: Harrison Edwards, Inc.
Typesetting: Trufont Typographers, Inc.
Printing: Eastern Press

front cover
Theodore Roszak
Man at Machine, 1937
oil on canvas, 24 x 40⅛″
The Regis Collection, Minneapolis, Minnesota

back cover
Nam June Paik
John Cage, 1990
mixed media sculpture with video, approx. 7′
Collection of Carl and Elizabeth Solway

frontispiece
Barbara Morgan
Wild Bee Honeycomb/Skyscraper, 1973
photomontage, 9½ x 6⅜″
Barbara Morgan

Artists in the Exhibition

Ida Abelman
John Atherton
Alice Aycock
John Baeder
Style of John Henry Belter
Henry Billings
Margaret Bourke-White
Anton Breuhl
Judith K. Brodsky
Bruce Charlesworth
Alvin Langdon Coburn
Harold Costain
Ralston Crawford
Lydia Dona
Orshi Drozdik
Thomas Eakins
William England
Walker Evans
James Evanson
Phillip Evergood
John Fekner
R.M. Fischer
Larry Gipe
Waylande DeSantis Gregory
Jan Groover
O. Louis Guglielmi
Hananiah Harari
Addie Herder
Eva Hesse
Lewis Hine
Barry Holden
Winslow Homer
Tishan Hsu
George Inness
Yvonne Jacquette
Jasper Johns
David Kapp
Edward Kienholz
Barbara Kruger
Dorothea Lange
Ellen Levy
Donald Lipski
Louis Lozowick
Paul Ludick
Helen Lundeberg
Chris Macdonald
Medrie MacPhee
Joseph Maresca
Winsor McCay
Denny R. Moers
Barbara Morgan
Ray Mortenson
Walter Murch
Eadweard Muybridge
Forrest Myers
Cady Noland
Claes Oldenburg
Dennis Oppenheim
Timothy O'Sullivan
Tony Oursler
Nam June Paik
Robert Rauschenberg
Man Ray
Ron Rocco
John Rogers
James Rosenquist
Theodore Roszak
Mrs. Saunders
Charles Sheeler
Peter Stackpole
Ralph Steiner
Alfred Stieglitz
William Stone
Jessica Stockholder
Paul Strand
Karl Struss
Abraham Joel Tobias
George Tooker
H.C. Torrence
Walter Von Nessen
Abraham Walkowitz
Andy Warhol
Kem Weber
Dan Weiner
Edward Weston
Hannah Wilke
Alexander H. Wyant
Nina Yankowitz

Lenders to the Exhibition

Aperture Foundation, Inc., Paul Strand Archive

Brooke Alexander

The George Arents Research Library for Special Collections, Syracuse University

Art et Industrie

The Art Institute of Chicago

Thomas Barry Fine Arts

David Beitzel Gallery

Dick Belger

Karl Bornstein Gallery

Dr. Peter and Eileen Broido

Leo Castelli

Tom Cugliani Gallery

Phillipe Daverio Gallery

Sid Deutsch Gallery

Martin Edelston

James Evanson

Gemini G.E.L.

John Gibson Gallery

Barbara Gladstone Gallery

Timothy Greenfield-Sanders

The Nora Eccles Harrison Museum of Art, Utah State University

Addie Herder

Hirschl & Adler Galleries

The Historical Society of The Town of Greenwich

IBM Corporation

International Museum of Photography at George Eastman House

Ruth Kaufmann

Carlo Lamagna Gallery

Lewis Lehr

Ellen Levy

Library of Congress

Donald Lipski

Vera List

Joseph Maresca

James Maroney

The Metropolitan Museum of Art

Loretta Michaelcheck

Robert Miller Gallery

Mississippi State Historical Museum

The Museum of Modern Art

Ray Winsor Moniz

Montclair Art Museum

Barbara Morgan

Ray Mortenson

Alan Moss Studios

Forrest Myers

National Museum of American Art, Smithsonian Institution

National Museum of American History, Smithsonian Institution

New York Public Library

The Oakland Museum

Bjorn Olsson Gallery, Stockholm, Sweden

Mr. and Mrs. John Ong

Tony Oursler

Philadelphia Museum of Art

Photofind Gallery

Robert Rauschenberg

The Regis Collection

Ron Rocco

Naomi and Walter Rosenblum

San Francisco Museum of Modern Art

Holly Solomon Gallery

Carl and Elizabeth Solway

Susan Teller Gallery

Abraham Joel Tobias

George Waterman III

John Weber Gallery

Whitney Museum of American Art

Hannah Wilkie

Yale University Art Gallery

Nina Yankowitz and Barry Holden

Richard York Gallery

Zabriskie Gallery

Private Collections

Contents

Foreword

The Technological Muse, the inaugural exhibition for the new Katonah Museum of Art building, is a tribute to the late John I. H. Baur, Chairman of the Exhibition Advisory Board for over 30 years. His vision and guidance shaped the exhibition philosophy of the Katonah Museum of Art. Jack Bauer believed that museums offer tangible evidence of the the creative energy of civilization. "Art," he said, "is not a furbelow but a profound expression of man's spirit."

The Technological Muse seems to us, too, to be particularly timely. As we approach the beginning of a new century, it is fitting to pause for a moment, glance back at one of the most powerful forces that shaped American society, and contemplate the implications of that force on our future. Surely, *The Technological Muse* examines a persistent and constantly evolving issue in the continuum of American art history: the impact of technology and the machine on our society. American artists have viewed this phenomenon with adulation, pessimism, and ambivalence.

A word about our new building designed by Edward Larrabee Barnes . . . it is elegant, intimate in scale, and in complete harmony with its beautiful site. In many ways, the design is analogous to the institution itself. The Museum has long served a dual function as a valuable community education resource and regional arts institution. Similarly, the new building is designed to fit into the context of its neighborhood, yet provide the kind of flexible exhibition space and state-of-the-art technology necessary to mount important exhibitions.

It is our belief that *The Technological Muse* is the appropriate exhibition to inaugurate the new building, where ideas and objects are interwoven in an open process of communication that blends exploration and study; where the human heritage is presented in its many voices; where personal visions of that heritage are sparked; and where the individual can seek and discover a cultural context. In this environment, one develops the ability to synthesize ideas, form opinions, and shape an aesthetic and cultural sensibility.

My deep appreciation extends to Dr. Susan Fillin-Yeh, Guest Curator; and to Angela Gilchrist, Exhibition Committee Chairman, and Nancy Wallach, Docent Chairman, who served as Exhibition Coordinators. We are grateful to all the lenders who took part in this project.

The New York State Council on the Arts and the Council for the Arts in Westchester are to be thanked for their ongoing support. My warm thanks to the Inaugural Exhibition Patrons for their generous support and to the Ida and William Rosenthal Foundation for its leadership gift toward our Publications Fund.

George G. King

Director

Acknowledgments

While many individuals have contributed to this project, our deepest gratitude goes to Betty Himmel, Exhibition Chairman from 1976-1988, who provided the initial impetus that led us to *The Technological Muse*. Three years ago, the Exhibition Committee explored possible topics for the inaugural exhibition in the new building. These preliminary discussions coincided with the untimely death of John I. H. Baur, Chairman of the Exhibition Advisory Board of the Katonah Museum of Art (formerly The Katonah Gallery) and Director Emeritus of the Whitney Museum of American Art.

It was clear that the inaugural exhibition should be drawn from one of Jack Baur's ideas, thereby acknowledging his distinguished scholarship as well as his devotion to the Katonah Museum. Dr. Jules D. Prown of Yale University, a colleague of Baur's and fellow champion of 19th-century American art, recommended that we meet with Dr. Susan Fillin-Yeh. She was then working on material that developed a theme Jack Baur had begun to explore in the 1950s, namely, the ambivalence of American artists toward technology and the machine.

Dr. Fillin-Yeh's proposal for *The Technological Muse* was enthusiastically accepted by the Museum's Exhibition Committee in the spring of 1989. To Dr. Fillin-Yeh we extend our thanks and congratulations. Her original scholarship and collegial spirit have been exemplary. We also express our gratitude to Dr. Prown and Louisa Baur for their early encouragement and ongoing support.

An exhibition the size and scope of *The Technological Muse* requires the combined energies and talents of people both inside and outside the Museum. We are particularly grateful to all the collectors, public and private, who generously agreed to lend us their works.

Arthur Clark, who had the formidable task of designing the exhibition, created, with characteristic ingenuity, an installation that projects an inherent inevitability. He was ably assisted by Doug Kumins.

The Curator joins in extending a special thanks to the artists who opened their studios and assisted us in selecting exhibition material: Alice Aycock, John Baeder, James Evanson, Addie Herder, Yvonne Jacquette, David Kapp, Ellen Levy, Donald Lipski, Paul Ludick,

Medrie MacPhee, Ray Mortenson, Forrest Myers, Claes Oldenburg and Coosje Van Bruggen, Ron Rocco, and Hannah Wilke.

Dr. Fillin-Yeh owes a special measure of gratitude to Naomi and Walter Rosenblum, and to Writer's Group members Julia Ballerini, Wendy Fairey, Rita Kashner, Pat Mainardi, and Sally Steinberg for their valuable comments on the catalogue essay. Aline Benjamin, Barbara Folsum, and Richard Seidlitz thoughtfully read the manuscript.

And to all those individuals who were kind enough to share their time, knowledge, and expertise, we express deep appreciation: Susan Arthur, John Boone, Catherine Brawer, Leo Castelli, Tom Cugliani, Susan Danly, Colin de Land, Pat Ferrero, Howard Greenberg, Barbara Haskell, John Hanhart, Pat Hearne, Peter Liebhold, Susan Lubowsky, Mary Lucier, James Maroney, Mark Miller, Ray Moniz, Anthony Montoya, Sarah Morthland, Alan Moss, Barry Newman, Elizbeth O'Brien, David Platzker, Carole Pesner, Howard Read, Ray Ring, John Rogers, Steven Rosen, Carlene Stephens, Susan Teller, Julia Van Haaften, Mimi Wheeler, Mitchell White, and Virginia Zabriskie.

The Katonah Museum of Art has always called itself a teaching museum, and we acknowledge the efforts of all those who have made this exhibition an educational experience. Yvonne Pollack, Director of Education, and her associate Lisa Stewart have devised exciting activities for school classes, weekend events, and family programs. Myrna Clyman and Betty Horowitz have arranged a series of public programs that provide a deeper insight into some of the issues raised by this exhibition.

Our dedicated docent corps has worked long and hard to interpret the works and offer provocative and illuminating tours. Rosalie Dolmatch has prepared another in her series of informative audio tours.

The monumental job of keeping track of all the materials pertaining to this exhibition could not have been accomplished without the help of Julia March and members of the Museum staff—Pete Brocki, Edris Guest, Mary Brenda Joyner, Pat Keane, Barbara Plechaty, and Nancy Roth. Pete and Julia deserve special mention. Rosalie Dolmatch, Virginia Gold, Betty Himmel, Katherine Moore, and Carolyn Mandelker contributed their editorial finesse. Annabelle Gundlach, a valuable intern during the summers of 1989 and 1990, was also extremely helpful.

Finally, we thank our director, George King, who had his hand in every phase of the planning of the exhibition. He encouraged us from its inception, and advised and assisted us with countless details and decisions.

The Katonah Museum of Art produces exhibitions that are original, dynamic, and educationally relevant, and continues to create an atmosphere in which it is a distinct honor and delight to collaborate.

Angela H. Gilchrist, Nancy Wallach

Exhibition Coordinators

The Technological Muse: Affirmation and Ambivalence in American Machine Imagery, 1840–1990

The civilization which took form in the United States [after 1776] was . . . technological civilization.
John A. Kouwenhoven *The Arts in Modern Civilization*, 1948

To count upon technological advance for ***anything but*** *technological advance is a mistake.*
Ursula K. Le Guin *A Non-Euclidean View of California as a Cold Place to Be*, 1982

Ever since the Industrial Revolution, American painters, sculptors, artisans, and photographers have been creating images of machines, technology, and the industrial landscape. Technological change has been examined through visual statements that are provisional, conflicted, contradictory, subversive, and iconoclastic. *The Technological Muse* draws attention to these often problematic images by tracing their development from the nineteenth century to the present. The questions they raise comprise the theoretical spine of the exhibition.

The Technological Muse brings together objects each of which is bound up in its own history and social fabric. At the same time, certain thematic clusters emerge to provide vantage points for the exhibition: photography; the railroad in the landscape; the automobile; beautiful machines; impossible machines, anti-mechanical constructions, and robots; the industrial landscape and the ecology; factory life and the worker; the industrialized cityscape with its skyscrapers and bridges; the machine in the home.

These groupings represent successive working hypotheses about technological materials and processes. Some of them spiral into the future, for the arts do more than mirror society and events. In fact, visual culture gives shape to the world in attempting to make sense of it. The paintings, sculptures, photographs, and decorative arts in *The Technological Muse* are the stories we tell about ourselves, our fears, and our desires.[1] Every object in the exhibition tells about the powers and the perils inherent in technology and our reactions to them. Each work speaks of the impact of mechanization on our lives and our world. Viewers as well as artists shape the meanings of these images. In *The Technological Muse* there are many voices. One might even say that there are many muses.

The muses of Greek mythology, all women, were mute; in *The Technological Muse*, women speak out. One of the earliest objects in the exhibition is *Train Quilt Top*, 1884, (pl. 15). Its blocky patterning depicts locomotives and railroad cars shuttling ceaselessly on their tracks in every direction. Mrs. Saunders, its maker, embroidered a title on it: "The Age In Which We Live." Hers is a new synthesis, a vision of a radically transformed world reworked into the reassuring, homey form of a quilt; technology is absorbed into the fabric of domestic intimacy. Quotidian and unpretentious, Saunders' quilt suggests options beyond the conventional opposites of mechanical/organic, hard/soft, public/private.[2] Her thinking pays indirect tribute to Mary Shelley, whose bestseller, *Frankenstein; or, The Modern Prometheus (1818)*, offered an early humanist and feminist critique of the applications of technology. Shelley invented the character of mad scientist Victor Frankenstein, who, alienated from both human relationships and nature, created a monster, the first robot.[3] "I pursued nature to her hiding places,"[4] Frankenstein boasted. Following in a long tradition of gendered visions of power, Frankenstein defined nature as female; culture, science, and technology as male.[5]

Technological transformation tended to be represented in terms of sexual difference in the nineteenth century, as can be seen by comparing two closely related depictions of the same river valley. Thomas Cole's *View on the Catskill, Early Autumn*, 1837 (Fig. 1) was painted before the railroad was built and his *River in the Catskills*, 1843 (Fig. 2) was painted afterwards. In the earlier work, a young child near a tree by a brook extends her arms toward a woman picking wildflowers who is offering them to her. They are the foreground figures in a vast and lush green landscape that seems uninhabited, although there are scattered plumes of smoke from chimney fires in the distance. A horse runs in a meadow. In the later picture, the tree is down and its chopped limbs litter the foreground. A man, axe in hand, now stands by the creek; felling trees has replaced picking flowers. Cole set the log-

Figure 1
Thomas Cole
View on the Catskill, Early Autumn, 1837
oil on canvas, 39 x 63"
The Metropolitan Museum of Art, Gift in memory of Jonathan Sturges by his children, 1895

Figure 2
Thomas Cole
River in the Catskills, 1843
oil on canvas, 28¼ x 41¼"
Gift of Martha C. Karolik for the M. and M. Karolik Collection of American Paintings, 1815–1865, Courtesy Museum of Fine Arts, Boston

ger's figure directly in line with the small but dominant middleground image of an "iron horse," a locomotive spewing smoke as it moves across a recently built viaduct spanning Catskill Creek. New houses are seen. There are two children in a cleared field, but note that the woman has disappeared along with the primeval landscape with which Cole once identified her. She is literally out of the picture, removed by the artist from this scene of nascent technological dominance.

Cole's railroad paintings were early models for one of the nineteenth century's most popular subjects. The railroad, as a transforming agent in the landscape, appears again and again in art as a visual marker for the spread of technological civilization. Images of the railroad were also chilling reminders of the exploitation of the land and the shameless annihilation of Native Americans.[6] The image of a smoke-spitting locomotive became the nineteenth century's emblem of the new American landscape. It was not only the landscape that changed. With the advent of photography in the 1850s, techniques for representing the world changed as well.

Railroad imagery and the new art of photography are the starting points for the following brief history of technology in American culture and form the basis for a critique of technological change. "The young man who expects to succeed," wrote William Dean Howells in 1900, "must . . . have a snapshot style in conversation, and patronize rapid transit vehicles."[7] Add "and woman" to Howells' sentence, and the comment is a useful one. "A snapshot style" was appropriate for the new age.

Photography

Form is henceforth divorced from matter.
Oliver Wendell Holmes, 1859

Photographs–machine-made images–seemed miraculous to a nineteenth-century audience. A "triumph of human ingenuity . . . audacious, improbable, incredible . . . an inconceivable wonder," wrote Boston physician and amateur photographer Oliver Wendell Holmes in 1859.[8] Photographs, Holmes noted, seemed to have even more substance than their originals. This comment directs attention to a fundamental aspect of photography that we sometimes overlook. Although photographs are the objective products of a mechanical process, they also reveal the subjectivity of the person who snaps the shutter. Since all imagery is interpreted according to the experience of the viewer, meaning is ultimately relative. The diverse images in *The Technological Muse* are testimony to the versatility of the camera as a machine. Photography is a precise technology. But there is contingency at the heart of its precision. This shifting dynamic that runs counter to notions of mechanical predictability is a thread that winds through 150 years of art and photographic history. It is particularly evident in paintings with photographic referents, for example, Hananiah Harari's *Smiles*, 1940 (pl. 11).

John Roger's *The Photographer and the Sitter*, 1878 (pl. 9) is a sculpture in two pieces–one, the photographer and his camera, the other, his sitters. Like the photograph the photographer is taking, the sculpture itself could be mass-produced and was intended for a mass market. Modern technology has provided the subject, the process, and the materials for this work. There is an inherent contradiction, however, for the look of the plaster composition had a specific meaning for a nineteenth-century audience, alluding to classical sculpture.

But if, as the nineteenth century first saw it, the camera produces authenticity, then authenticity itself is a construct, as can be seen, for example, in Timothy O'Sullivan's photographs. *Destruction of a Railroad Bridge*, 1864 (pl. 4) showed its viewers a face of war with which they were just then becoming familiar, for the Civil War, America's first highly mechanized war, was also her first photographed one. O'Sullivan's photographs gave a new and nearly unbearable vision of events never before seen in their immediacy–wholesale destruction and death in combat. These shocking realities were difficult to accept. When published, as they were, in albums, the impact of the photographs was modified by allegorical or moralizing texts.[9] Public sculpture commissions commemorating the Civil War did not take these photographs as their models; rather, they drew on older visual conventions for heroism.

Photographs of the Western landscape had a similar aura of unreality for nineteenth-century viewers. O'Sullivan's *Devil's Slide* series, c. 1870 (pls. 5, 6, 7, 8) entered the public domain as fantastic imagery. Even today, they seem almost overcharged with a dense layering of detail—"form," as Holmes had suggested, divorced from "matter."

In the late nineteenth century, time-lapse photographs were made possible by electrifying shutter mechanisms. The first of these were Eadweard Muybridge's motion studies, among them *Head Spring, a flying pigeon interfering*, 1874 (pl. 1). Muybridge's pictures provided new information about the sequence of human and animal movement. Thomas Eakins' bronze sculptures of trotting horses (pl. 2) were directly based on Muybridge's studies. Both sculptures and photographs showed familiar forms frozen in unfamiliar positions. *Head Spring* is the unique record of the planned and the accidental movements of a falling man who catches himself.

It is no accident that Winslow Homer's engraving of a Civil War scene, *The Army of the Potomac, A Sharpshooter on Picket Duty*, 1862, (pl. 3), seems to anticipate a frozen time segment from one of Muybridge's photographs. Such visions had been in the air even before Etienne Jules Marey's photographic gun (1882) brought the phrase "shooting a photograph" into use.

The Railroad in the Landscape

Railroad carriages are pulled . . . through the landscape setting fire to crops, scaring the livestock and frightening women and children. The Almighty certainly never intended that people should move at such breakneck speed.
Martin Van Buren Governor of New York, 1829

Images of the railroad in the landscape illustrate the shifting balance between pastoral and industrial concerns in nineteenth-century literature, described by Leo Marx in his book *The Machine in the Garden*.[10] George Inness' *Delaware Water Gap*, 1857 (pl. 14) shows a railroad train running through cleared and cultivated fields, which he would have referred to as a civilized landscape.[11] Such paintings, depicting a temporary balance between Nature and Technology, served as advertisements for the railroad and were often reproduced and widely distributed as Currier and Ives' lithographs. The railroad was a popular subject for photographers as well as painters. Photographers, however, tended to place technological innovation in the foreground by monumentalizing trains, trestles, and railroad bridges.[12] In William England's *Niagara Suspension Bridge*, 1859 (pl. 12) Niagara Falls, nineteenth-century America's pre-eminent tourist attraction and national icon, has been displaced from the foreground by William Roebling's suspension bridge built in 1855 and by the train moving across it.

In the twentieth century, railroad imagery developed new associations. During the 1920s, trains were portrayed as glamorous, streamlined forms reflecting the popular interest in up-to-date design as seen in Henry Dreyfuss' photograph of the New York Central Railroad-Twentieth Century Limited (Fig. 3). Later during the Depression years, the train became an ironic symbol of dislocation and poverty, used tellingly in socially concerned photographs for the Farm Security Administration. Dorothea Lange, 1936 (pl. 13) focuses on

Figure 3
Henry Dreyfuss
[New York Central Railroad—Twentieth Century Ltd., no date]
photograph, 8½ x 10″
From the Leslie Merrill Collection in the Huntington Library, San Marino, California

homeless people who could not afford to ride the trains. In 1950, George Tooker used images of subway riders to portray psychological displacement (pl. 16). More recent paintings bear witness to the near disappearance of rail travel and reflect nostalgia for a bygone era. John Baeder's *Maple Diner*, 1973 (pl. 17) looks like a railroad car taken from its tracks.

The Automobile

ONE DIRECTION! whoops America, and sets off . . . in an automobile.
D.H. Lawrence, 1923

In the twentieth century, the automobile displaced the railroad as the symbol of technological development. A new "machine age" vision of beauty can be seen as early as 1917 in Dadaist Francis Picabia's unconventional portrait of his wife, Gabrielle Buffet, as a sports-car's windshield. (Picabia felt Buffet was his faithful protector.) Margaret Bourke-White, in her photograph, *La Salle*, c. 1935 (pl. 18) saw beauty in the precisely equilibrated curves of a gleaming metal fender. Some sixty years of advertisements have continued to connect the automobile with notions of romance and power. But automobiles in paintings, sculpture, and photographs do not necessarily support this interpretation. Often they are unsettling images, hinting at violence. In Dorothea Lange's Farm Security Administration photograph, *Plantation Owner and His Field Hands, Mississippi Delta*, 1936 (pl. 21) the automobile connotes economic and political domination. The white plantation owner with his foot on the bumper of his car shows us that he owns and controls it, just as he owns and controls his field hands. A similar proprietary gesture of power is enacted in Dan Weiner's later photograph (pl. 19) in which the dialectic is a sexual power play, for the swelling curves of the 1950s Cadillac can be seen as references to the female body. The shabby, stranded car in John Atherton's *Christmas Eve*, 1941 (pl. 23) is anthropomorphized as a Surrealist-influenced metaphor for the human condition. Closer to our times, images of automobiles are often threatening. Robert Rosenquist's photo-realist *Untitled (Broom Street Truck)*, 1965 (pl. 22) looms dangerously close. It is interesting to juxtapose this work with Chris Macdonald's toylike *Work Truck Variation*, 1989 (pl. 40) in which scale is also exploded. David Kapp's *Oncoming Car X*, 1989 (pl. 20) enters our space; we see it with the eyes of someone about to be run over by it. John Fekner, whose grafitti art, like Barbara Kruger's, occupies urban spaces, stencils deadpan messages on deteriorating walls, highway and bridge embankments—and on junked vehicles. *Industrial Fossil*, 1978, 1990 (pl. 24) is one reality of automobile culture.

Beautiful Machines

Who could expect . . . to see beauty in a thing not made for beauty: the Machine.
Jane Heap, 1927

Nineteenth-century machinery conforms to an artistic sensibility that now seems remote. A *Hawkins Mitering Machine*, c. 1870s (pl. 33) with its elaborate floral embellishment, seems anti-technological to twentieth-century eyes.[13] The emergence of a "machine aesthetic" in the early twentieth century brought with it an appreciation of undecorated, polished metallic surfaces, sharp-edged geometric forms, and smoothly meshing gears.[14] Functional objects were recognized as beautiful and became the vocabulary for a new art (pls. 31, 32). Paul Strand's photographic experiments with abstract forms of machinery are considered icons of modernist photography (*Gear, New York*, c. 1920, pl. 27).

Artists in the 1930s and 40s, among them Henry Billings (pl. 25), Lewis Hine (pl. 76), and Ralph Steiner (pl. 26), often juxtaposed people and machinery to suggest relationships of comfort and respect. Ralph Steiner places Louis Lozowick, a Precisionist painter of machinery and factoryscapes, in front of a milling machine so large that it barely leaves room for the sitter. Lozowick, thoughtful, is at ease. Steiner positions his camera to emphasize the complex meshing of interconnecting shapes: Lozowick's head, his intelligent sharp features, his eyeglasses, the pronounced whorls and shadows of his ear, the gears behind him. These gears suggest the smooth functioning of the human mind—clean, oiled, in good working order. Thus, the attributes of the machine have been transferred to humanity.

Conversely, artists began to attribute human qualities to machines. In paintings by Helen Lundeberg, *The Mirror: Enigma*, 1934 (pl. 29) and Ida Abelman, *My Father Remi-*

nisces, 1937 (pl. 28) products of technology evoke memory, nostalgia, mystery—non-technological qualities. Walter Murch expresses a haunting "peculiar beauty"[15] in Surrealist-influenced depictions of machinery. He reveals the inner workings of *The Calculator*, 1949 (pl. 30): one side has been removed, and plugs and light bulbs lie on the ground next to it. His precise rendition is a portrait of the machine as romantic hero, clad in a cape-like carapace brooding against a lowering sky.

More recent works depict a technology that has become increasingly dysfunctional. Joe Maresca's *The Fruits of Our Labor*, 1987 (pl. 34) improbably combines gears, drill bits, Greek columns, lush fruit, and jewelry. Cady Noland's *Cart Full of Action*, 1986 (pl. 41) a metaphor for technological consumerism, is a heavy-duty industrial cart, filled with automobile hubcaps and gleaming machine parts of the sort first considered beautiful in the 1920s. But they do not fit together or function as a mechanism; they are piled any which way, and the cart must be pushed by hand. In this respect, Noland's is an impossible machine.

Impossible Machines, Anti-mechanical Constructions, Robots

People living in a machine age are naturally influenced . . . by the age they live in.
I introduced derision into that sacrosanct era.
Marcel Duchamp, 1960

In the early 1920s in New York City, the Dadaists, a loosely linked circle of avant-garde artists, writers, and musicians, followed the lead of Francis Picabia, Marcel Duchamp, and Man Ray, taking a particular interest in machinery (see also pl. 42). "I intend to go on working until I attain the pinnacle of mechanical symbolism," Picabia told bewildered newspapermen on his arrival from France.[16] Duchamp's particular legacy was contradiction and impossibility. Man Ray's *Object Without Title or Date*, c. 1921 (pl. 43) was originally an axle. *Object* simultaneously undercuts notions about art and technology. The piece is a "readymade," a term coined by Duchamp to describe common objects defined as works of art. Man Ray's axle contradicts its original function as a smoothly turning wheel base. If touched, *Object* moves with a lurching, dipping motion, rotating on its axis. It goes nowhere. *Object* is sexual, playful, a little nasty.

Duchamp's *Large Glass,* or *The Bride Stripped Bare By Her Bachelors, Even . . .* , 1915-23 (Fig. 4) was the signature piece of New York Dada. This vision of robotic mechanomorphs and mechanical sex (an enormous construction of metal, wire, lead paint, and even dust, mounted on glass) spawned generations of mechanomorphs. Hannah Wilke's video reconstruction of Duchamp's *A Bride Stripped Bare*, 1976 (pl. 44) Tishan Hsu's *Body Scanner #1*, 1988 (pl. 38) and Nam June Paik's *John Cage*, 1990 (back cover) belong to this genre, particularly since in the 1950s and 60s John Cage learned from Duchamp to mix precision with chance. Duchamp's presence in New York in the 1950s and 60s encouraged

Figure 4
Marcel Duchamp
Large Glass or *The Bride Stripped Bare by Her Bachelors Even*, 1915-23
oil and lead wire on glass, 109¼ x 69⅛"
Philadelphia Museum of Art, Bequest of Katherine S. Dreier

Pop artists, whose work questioned the value of technological expansion in the aftermath of the atomic bomb, the Cold War, and America's new prosperity.

Addie Herder's collaged homage, *A Machine for Luddites*, 1983 (pl. 54) and Edward Kienholz's *The Friendly Grey Computer—Star Gauge Model 54*, 1965 (pl. 39) reflect ambivalence. Kienholz's work has become an icon of contemporary folklore. A humanoid rocking chair with face-like control panel and doll's legs—this fabled "brain" presents no threat. "You can tell when it's . . . in a funky mood."[17] Jasper John's non-functional bronze *Light Bulb*, 1960 (pl. 47) questions assumptions of both art and utility. Other Pop artists such as Robert Rauschenberg (*Jellyfish Traffic*, 1986, pl. 48) and Claes Oldenburg (*Ice Bag—Scale B* 1971, pl. 46) subvert the functional aspect of ordinary objects. The *Ice Bag* was produced under the auspices of Maurice Tuchman's Art and Technology program at the Los Angeles County Museum, which helped to reframe conventional notions of technological production in the 1960s.[18] Oldenburg wrote that he wanted to "remove the difficulty of technology . . . and make it do some very foolish thing."[19] Eva Hesse's *Untitled*, 1965 (pl. 53) distorts and twists industrial materials and is an early example of her concern with extremes and absurdities. The central column and scythelike cutting blades of Alice Aycock's *DNA Cutter*, 1984 (pl. 51) evoke both the spiral helix of the ancient Mediterranean Mother Goddess religion and the structure of the DNA molecule (see also pl. 50). Orshi Drosdik's *The Black Mirror*, 1988 (pl. 37) is a photograph of a horrific nineteenth-century museum exhibit, a child's head preserved in formaldehyde. The serene expression on its face compels us to reconsider the goals of scientific technology. Barbara Kruger reminds us, "*Your Manias Become Science*." 1981 (pl. 35).

Threatening machines may be considered a subset of impossible mechanisms (see also pl. 52), which function only in the mind. Bruce Charlesworth's *25 Fate*, 1986 (pl. 36) portrays in lurid light his shocked response to something we cannot see on a television screen—something dangerous, something "really alive," something that has drawn him too close. In addition to portraying the machine as a dangerous mechanism, twentieth-century artists also see the machine as a powerless agent, for example Lewis Hine's *Junkies Breaking Up Old Looms for Scrap, Paterson, New Jersey*, 1937 (pl. 49). Other artists have sought a cooler, more aesthetic dismemberment. Jan Groover's *Untitled [Objects with Tube & Circle]*, 1986 (pl. 55) treats individual machine parts with geometric, Platonic balance, visually removing them from any connection with their function (see also pl. 56).

The Industrial Landscape and the Ecology

The garbage mountains exploding . . .
Judith K. Brodsky, 1989

The harsh realities of the nineteenth-century factoryscape were often treated in the softly focused, romantic style of the period (pl. 70). Better known are those images produced in the 1920s, 30s, and 40s by painters and photographers we now call the Precisionists. These artists saw echoes of Greek and Roman classicism in the cubistic forms of the modern industrial landscape. The clean, balanced shapes of Edward Weston's *Pipes and Stacks (Armco Steel, Ohio)*, 1922 (pl. 61) are based on ideal geometric proportions. The sharply-focused photographic sensibility that is in evidence here (see also pls. 71, 72) is characteristic of the Precisionist "machine style" and was used in painting as well as in photography.[20] In fact, Charles Sheeler (pl. 79) and Ralston Crawford (pls. 66, 67) painted industrial subjects directly from their own photographs. Factoryscapes of the 1920s rarely included human subjects. In the 1930s, this omission took on a new meaning: deserted factories became visual metaphors for the social and economic disasters of the Great Depression. When, in 1935, Walker Evans photographed *View of Easton, Pennsylvania on the Lehigh River* (pl. 69) for the Farm Security Administration, he looked down on the town to produce a laminate of dual, conflicting messages. Photographic distancing presented a view of town and factory neatly separated by a river, editing out Easton's population. In Charles Sheeler's conté crayon drawing *Industrial Architecture*, 1931 (pl. 79) the building dominates and dwarfs the head of the man in the lower right. In Louis Lozowick's *Dusk*, 1932 (pl. 78) a sleeping man, presumably homeless and unemployed, lies next to a deserted factory, illustrating the marginal position of workers during the Depression years.

Like uninhabited contemporary factoryscapes, Denny Moer's *Figment from a Landscape*, 1989 (pl. 64) is profoundly inhospitable. It is even dangerous for humans to be there. Larry Gipe inscribed the word "Faith" in large gold letters across the bottom of his haunting, atmospheric night scene, *Untitled 5 from the Factory Series*, 1987 (pl. 62) to remind the viewer that words are necessary; technology's benefits are no longer evident. The origins for such powerful images can be traced back to paintings such as Sheeler's *Conversation—Sky and Earth*, 1940 (pl. 63; see also pls. 58, 68). The hydroelectric installation dominates the horizon just as its technology controls nature's resources. Although he may not have realized it, Sheeler was prescient. In other works, the industrial landscape is not only uninhabited but also unsafe. Industrial waste, the by-product of a technological society, is the subject of Judith Brodsky's *Garbage Bag Mountain*, 1989 (pl. 59). Ray Mortenson's photograph, *Newark, Harrison, Amtrak Bridge from the South Bank of the Passaic River*, 1982 (pl. 65) is beautiful until one notices the garbage floating in the polluted water. Tony Oursler's *Psychomimetiscape*, 1988 (pl. 73) juxtaposes the cooling stacks of a nuclear power plant with the tower of a medieval castle. A videotape, enclosed in the tower, plays the conversation of nuclear plant workers, undercutting any implication of romance or adventure.

Factory Life and the Worker

Factory windows are always broken . . .
Vachel Linsday, 1914

Images of factory life have often been questioning, unsettling, even distrustful. One of the earliest, Winslow Homer's *Morning Bell*, 1866 (Fig. 5) shows a young woman walking briskly up a footbridge on her way to work. The footbridge is makeshift and rickety. The factory windows are dark, some of their panes missing. And although the woman's fashionable jacket (purchased, presumably, with factory wages) is as bright as the morning sunlight, her attire sets her apart from a group of women at the right in homespun clothes. We may interpret this painting as a premonition of the failure or betrayal of the industrial dream.

Twentieth-century artists continued to be concerned with the impact of the factory on the lives of its workers (pls. 74, 75). Lewis Hine's *West Virginia: Small Boys Worked All Night in the Glass Factories*, June, 1911 (pl. 82) is a poignant commentary on the exploitation of child labor. Hine presents a more heroic image, a wishful counter to reality, in the

Figure 5
Winslow Homer
The Morning Bell, c. 1866
oil on canvas, 24 x 38¼"
Yale University Art Gallery
Bequest of Stephen Carlton Clark, BA 1903

proud young worker of *Heart of the Turbine*, 1930 (pl. 76). A product of the Depression Era, it reaffirms the benefits of technology's promise of work and future paychecks. Waylande Gregory's *Farmer and Factory Worker* (pl. 77), made for the 1939 New York World's Fair (whose theme was prosperity through technology), is a ceramic statue representing the male worker, seemingly "muscle-bound in a symbolic trance."[21] His Janus-like figure seems awkward and ungainly, reflecting inappropriate classical references. By contrast, Theodore Roszak's *Man at Machine*, 1937 (cover) is a vital presence, competent, if somewhat frantic. Are we looking at a person or a machine? "Man becomes a robot . . ." reported Abraham Joel Tobias describing the factory in his *Machine Design*, 1934 (pl. 80), "so the figure in the foreground has no features."[22] Puppet-like, the apathetic people in O. Louis Guglielmi's *Relief Blues*, 1934 (pl. 81) are dulled by joblessness.

The Industrialized Cityscape: Skyscrapers and Bridges

Since last spring, Babylon has entered a new fever, the building fever. There is not a block where not a new skyscraper is popping up overnight like an asparagus. You can imagine the noise of mechanical hammering . . .
Stephan Bourgeois, New York Art Dealer, 1920

After 1916, when New York's "set back" zoning gave impetus to steel-scaffolding construction, skyscrapers became the icons of the American City.[23] Stephen Bourgeois' asparagus metaphor makes us aware of the speed with which the American city was changing, and, at the same time, suggests the reactions of his contemporaries to this transformation.

Abraham Walkowitz's abstract *Cityscape*, c. 1938 (pl. 86) implies the menace of towering skyscrapers that seem to be swaying, and also hints at a frenzied crowd below (see also pl. 85). These tipped images combine the exaggerated perspective torques of Cubist painting with the distortion of camera optics to give visual shape to physical sensation—the ver-

Figure 6
Advertisement from *Fortune* Magazine
June, 1940, page 29
Sponsored by Electro Metallurgical Company, a unit of Union Carbide and Carbon Corporation

tigo that accompanies the backward tilt of the head when looking up at a skyscraper. An advertisement from *Fortune* magazine, July, 1940 (Fig. 6) reinforces this interpretation of skewed urban architecture. The rakish angling of skyscrapers in the background, the prows of ocean liners in the foreground, the factory buildings, and even the streamlined kitchen at the right, have all, improbably, been tipped. This vision, which is one of a city seen as an activity or as an experience rather than documented as a place, was also adapted for bridge imagery, for example, in Peter Stackpole's *Abstraction From Under 40 Tons*, 1934 (pl. 89). Yvonne Jacquette's *Reflected Facade (White Hall Street) II*, 1987 (pl. 90) with its off-register plate glass, creates a witty visual metaphor of the uncertainties of urban experience. In Barbara Morgan's *Wild Bee Honeycomb/Skyscrapers*, 1973 (frontispiece) the city is, literally, a hive. (These twentieth-century expressions contrast with turn-of-the-century photographic images by Karl Struss (pl. 91) and Alvin Langdon Coburn (pl. 83) in which the urban scene is softly focused with Old Master tonalities.)

Machinery Domesticated: The Machine Enters the Home

Furniture tells all . . . The Wassily armchair (1926) is an easy chair in which one cannot relax for more than thirty minutes at a time.
Withold Rybcznski, 1987

Withold Rybcznski's comment on domestic comfort offers a revealing insight into the influence of technology on modern furniture design. For if, as 1920s critic Lewis Mumford believed, there was hope for "genuine modernism making its way from the bathroom and the kitchen to the rest of the house,"[24] modern furniture was experimental both in its construction (pl. 93) and as an expression of functional aesthetics. A Museum of Modern Art exhibition in 1934 failed to establish a functionalist definition of "machine art."[25] Walter Von Nessen's brushed aluminum *Standing Floor Lamp*, c. 1928 (pl. 100) looks like a piece of machinery, but has a torch shape that alludes to pre-industrial lighting. Kem Weber's *Skyscraper Vanity and Chair*, 1929 (pl. 92) is also a contradiction: functional domestic furniture in the form of urban architecture (see also pls. 87, 88). Incongruities are also seen in even the earliest examples of machine-made furniture such as Belter-style pieces, 1858-68 (pl. 94) where mechanical methods were used to achieve an ornate hand-crafted look.

In recent post-Modern work, furniture and (dis)comfort are on the way to becoming metaphoric: Paul Ludick's *Arms Chair*, 1986 (pl. 95), Robert Fischer's *Desk Set*, 1987-88 (pl. 97), and Nam June Paik's *Art Deco Clock*, 1989 (pl. 99) re-use precisely machined materials in subversive, anti-functional, and anti-technological ways (see also pls. 101, 102). Ludick's *Arms Chair* recycles handguns. Fischer's *Desk Set* appropriates a soup pot as a lamp shade, in a biting, double-edged commentary on the power play symbolism of a furniture icon, the executive desk. William Stone's *Fall Line*, 1989 (pl. 98) refers to "this country's first industrial cities . . . mill towns."[26] It takes the form of a nineteenth-century Shaker work desk, but it is an absurd, anti-functional object with water gurgling through it. And Jessica Stockholder's jerry-rigged construction *Wood, Lamps, Paint, Sheet Rock, Small Coffee Table*, 1987 (pl. 96) does not even function as furniture, although it appropriates beat-up chairs and tables.

Conclusion: Technology as Metaphor (the Future)

We all recognize the technological voice coming out of a machine.
Christa Wolf, 1989

Stockholder's art is instructive. Along with Ron Rocco's grass and neon construction, *The Waterline Project*, 1989 (pl. 57) and Donald Lipski's recycled steel buoy and dice, *Bad Buoy*, 1988 (pl. 45), it foreshadows other novel uses of technological materials. The years to come will revise and reshape our ideas about art and technology. Experiments in holography, in laser and computer-generated images, will lead to visual forms not yet imagined. But if, as Christa Wolf reminds us, our fascination with technology is a blind spot at war with our instinct for self-preservation and our humanity, we need such paradigms.[27] The technological muse may yet startle us into a more thoughtful consideration of aesthetic and technological issues not yet fully explored.

Susan Fillin-Yeh
Guest Curator

Notes

1. Carolyn G. Heilbrun, *Writing a Woman's Life*, New York, W. W. Norton, 1988.

2. See Pat Ferrero, Elaine Hedges and Julie Silber, *Hearts and Hands: The Influence of Women and Quilts on American Society*, San Francisco, The Quilt Digest Press, 1987, p. 11.

3. Anne K. Mellor, *Mary Shelley, Her Life, Her Fiction, Her Monsters*, New York, Methuen, 1988, p. 109.

4. *Ibid.*, pp. 107, 115.

5. These connotations, as Roland Barthes and other theorists of culture have pointed out, are so pervasive that we find them difficult to distinguish; we tend to look past them. See Roland Barthes, ("Myth Today," *Mythologies*, New York, Hill and Wang, 1972 pp. 109-58); see also Antonio Gramsci, *Selections from the Prison Notebooks*, edited and translated by Quintin Hoare and Geoffrey Nowell Smith, London, Lawrence and Wishard, 1971, pp. 294-97.

6. For a discussion see Susan Danly and Leo Marx, editors, *The Railroad in American Art: Representations of Technological Change*, Cambridge, Massachusetts, The Massachusetts Institute of Technology Press, 1988, pp. 1-51; 71-91, and *passism*.

7. William Dean Howells, as quoted in John Silllgoe, *Metropolitan Corridor: Railroads and the American Scene*, New Haven, Yale University Press, 1983, p. 23.

8. Oliver Wendell Holmes, "The Stereoscope and the Stereograph," *The Atlantic Monthly* 3 (June 1859): pp. 738-48, reprinted in Beaumont Newhall, editor, *Photography: Essays and Images*, New York, The Museum of Modern Art, 1980, pp. 53, 60.

9. Alan Trachtenberg, *Reading American Photographs: Images as History, Matthew Brady to Walker Evans*, Hill and Wang, 1989, p. 93.

10. Leo Marx, *The Machine and the Garden: Technology and the Pastoral Ideal in America*, New York, Oxford University Press, 1964.

11. George Innes, as quoted in Danly and Marx, *op. cit.*, p. 87.

12. *Ibid.*, p. 13.

13. See John Kasson, *Civilizing the Machine: Technology and Republican Virtues in America*, New York, Grossman, 1976.

14. For a specific discussion, see Barbara Zabel, "The Machine as Metaphor, Model and Microcosm: Technology in American Art, 1915-1930," *Arts* 57 (December 1982): pp. 100-105; Susan Fillin-Yeh, "Charles Sheeler and the Machine Age," The Graduate Center, C.U.N.Y., 1981. See also, Reyner Banham, *Theory and Design in the First Machine Age*, New York, Praeger, 1967; Siegfried Giedion, *Mechanization Takes Command*, New York, W. W. Norton, 1969; and Richard Guy Wilson, Dianne H. Pilgrim and Dickran Tashjian, *The Machine Age In America*, The Brooklyn Museum in association with Harry N. Abrams, 1986.

15. Walter Murch, as quoted in Judy Collischan Van Wagner, *Walter Murch*, Whitney Museum of American Art at Phillip Morris, New York, 1986.

16. Francis Picabia as quoted in K. G. Pontus Hulten, *The Machine as seen at the end of the machine age*, New York, The Museum of Modern Art, 1968, p. 83.

17. Kienholz, as quoted in Hulten, *op. cit.*, p. 190.

18. Among these were Swiss artist Jean Tinguely's *Hommage to New York*, a machine performance which self-destructed in the Museum of Modern Art Sculpture Garden, New York, on March 17, 1960, and the Experiments in Art and Technology Project at the Museum of Modern Art, New York, 1967-68. See Hulten, *op. cit.* pp. 166-75; 198-207, and also *A Report on the Art and Technology Program of the Los Angeles County Museum of Art*, Los Angeles County Museum of Art, 1971.

19. Oldenburg, as quoted in *A Report on the Art and Technology Program of the Los Angeles County Museum of Art*, Los Angeles County Museum of Art, 1971, p. 265.

20. See Karen Tsujimoto, *Images of America: Precisionist Painting and Modern Photography*, published for the San Francisco Museum of Modern Art by the University of Washington Press, Seattle, 1982., and also Susan Fillin-Yeh, Charles Sheeler's *Rolling Power*, in Danly and Marx, *op. cit.*, pp. 147-64.

21. For a discussion of the period, see Richard D. McKinzie, *The New Deal for Artists*, Princeton, New Jersey, Princeton University Press, 1973.

22. Tobias, as quoted in Jeffrey Wechsler, *Abraham Joel Tobias*, Jane Voorhees Zimmerli Art Museum, Rutgers, The State University of New Jersey, 1988, p. 21.

23. Irving Pond, "Zoning and the Architecture of High Buildings," *Architectural Forum* 35 (October 1921): pp. 131-34.

24. Lewis Mumford, "Modernism for Sale," *American Mercury* 16 (January-April 1929): p. 455.

25. For a discussion, see Sidney Lawrence, "Clean Machines at the Modern," *Art in America* 72 (February 1984): pp. 127-41; 166-68.

26. William Stone, "Fall Line," (Artist's Statement), 1989.

27. Christa Wolf, *Accident*, New York, Farrar, Straus and Giroux, 1989, *passim*.

Plates

Photography

plate 1

Eadweard Muybridge

"Head Spring, a flying pigeon interfering," from *Animal Locomotion*, vol. 5, plate 365, 1887

photograph, Image size: 8¼ x 14½", Sheet size: 18¼ x 23"

The Miriam and Ira D. Wallach Division of Art, Prints and Photographs, The New York Public Library, Astor, Lenox and Tilden Foundations

plate 2

Thomas Eakins

Models of Horses for "The Fairman Rogers Four-In-Hand," 1879 (cast 1946)

bronze

Left Leader Horse, length 12½"

Right Leader Horse, length 11⅞"

Left Wheeler Horse, length 11¾"

Right Wheeler Horse, length 11¹⁵⁄₁₆"

Phildelphia Museum of Art. Bequest of Mr. and Mrs. William M. Elkins

plate 3

Winslow Homer

The Army of the Potomac, A Sharpshooter on Picket Duty, 1862

wood engraving, 9⅛ x 13¾"

Private Collection

plate 6
Timothy O'Sullivan
Devil's Slide, Weber Canyon, Utah, c.1870
albumen print, 10⅝ x 7⅞"
Library of Congress

plate 7
Timothy O'Sullivan
Devil's Slide, Weber Canyon, Utah, c. 1870
albumen print, 7⅞ x 10⅝"
Library of Congress

plate 4
Timothy O'Sullivan
Destruction of a Railroad Bridge, (North Anna River, May 25, 1864)
albumen print, stereopticon view, 3 x 7"
Library of Congress

plate 5
Timothy O'Sullivan
Devil's Slide Bridge, c. 1870
albumen print, 7 15/16 x 10⅝"
Library of Congress

plate 8
Timothy O'Sullivan
Devil's Slide, Weber Canyon, Utah, c. 1870
albumen print, $7^{14}/_{16}$ x $10^{5}/_{8}$"
Library of Congress

plate 9

John Rogers

The Photographer and the Sitter, 1878

painted plaster

The Photographer: 18½ x 9½", The Sitter: 17 x 8½"

Courtesy of The Historical Society of the Town of Greenwich

plate 10
Barbara Morgan
Valerie Bettis Desperate Heart–II, 1944
photograph, 8½ x 6¹¹⁄₁₆″
Barbara Morgan

plate 11
Hananiah Harari
Smiles, 1940
oil on canvas, 10 x 11″
Richard York Gallery, New York

The Railroad in the Landscape

plate 12
William England
Niagara Suspension Bridge, 1859
albumen-silver print, 9½ x 11½"
Collection, The Museum of Modern Art, New York. Purchase.

plate 13
Dorothea Lange
Next Time Try the Train, Relax, c. 1935
photograph, 10¾ x 13¾"
Photofind Gallery, New York

plate 14
George Inness
Delaware Water Gap, 1857
oil on canvas, 32 x 52¼"
The Monclair Art Museum, Gift of Mrs. F.G. Herman Fayen

plate 15

Mrs. Saunders

Train Quilt Top, c. 1884

pieced and appliqued cotton, 101¼ x 90¼"

Collection of Mississippi State Historical Museum

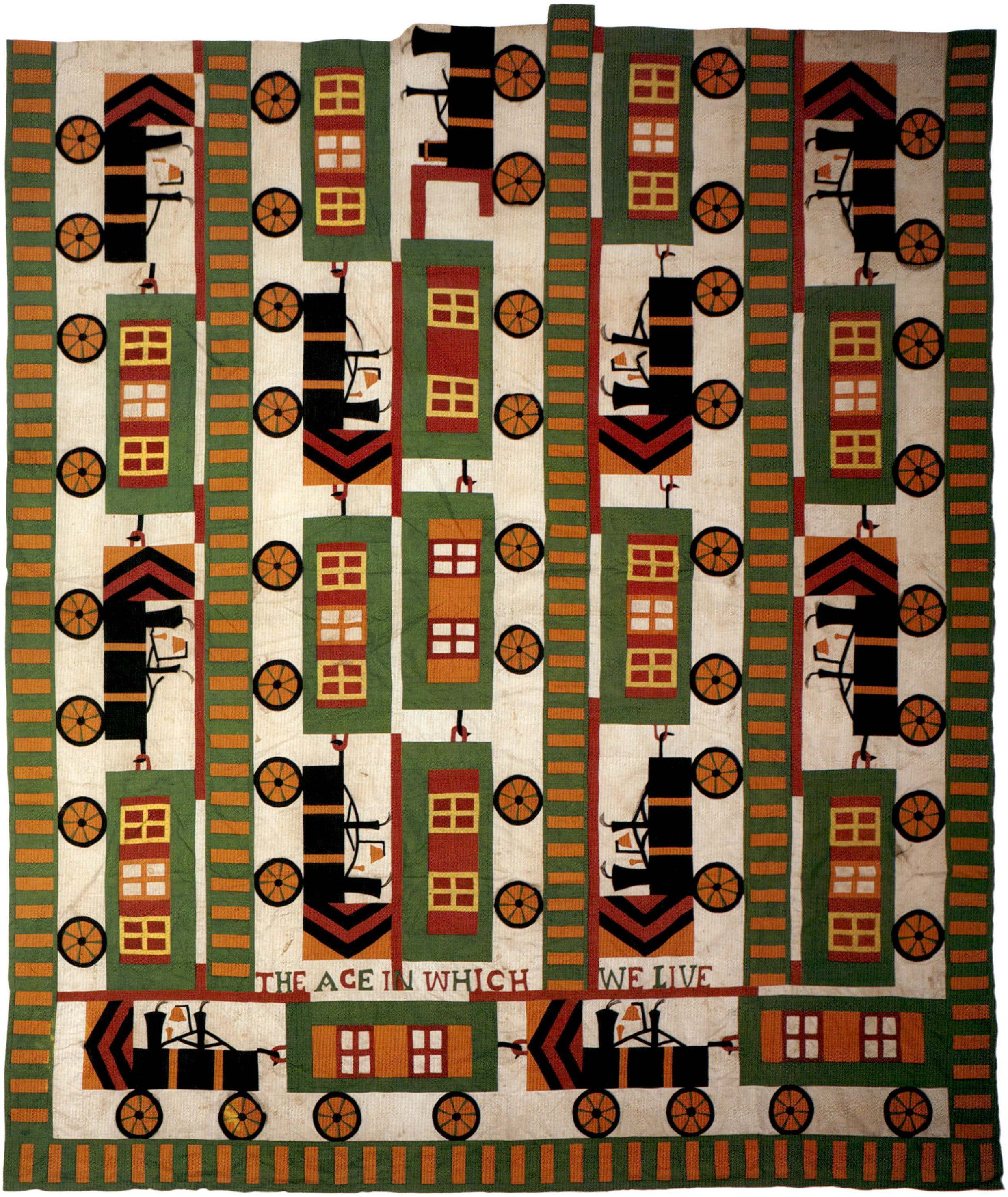

plate 16
George Tooker
The Subway, 1950
egg tempera on composition board, 18 x 36″
Collection of Whitney Museum of American Art.
Purchase, with funds from the Juliana Force
Purchase Award. 50.23

plate 17
John Baeder
Maple Diner, 1973
oil on canvas, 42 x 66″
Dick Belger Collection, Courtesy Morgan Gallery,
Kansas City

The Automobile

plate 18

Margaret Bourke-White

La Salle, c. 1935

photograph, 13⁵⁄₁₆ x 9⅜″

George Arents Research Library for Special Collections at Syracuse University

plate 19

Dan Weiner

Ross Sams, Church Furniture Maker, Waco, Texas, 1955

photograph, 9 x 13½″

Private Collection

plate 20

David Kapp

Oncoming Car X, 1989

oil on linen, 24 x 48″

Private collection, Courtesy David Beitzel Gallery, New York

plate 21

Dorothea Lange

Plantation Owner and His Field Hands, Mississippi Delta, 1936

photograph, 10 x 8″

Private Collection

plate 22

James Rosenquist

Untitled (Broome Street Truck), 1963

oil on canvas, 72 x 72"

Collection of Whitney Museum of American Art.
Purchase 64.20

plate 23
John Atherton
Christmas Eve, 1941
oil on canvas, 30¼ x 35″
Collection, The Museum of Modern Art, New York. Purchase.

plate 24
John Fekner
Industrial Fossil, 1978, 1990
Reenacted site-specific installation originally spray painted on truck—Maspeth, N.Y.
Courtesy Bjorn Olsson Gallery, Stockholm, Sweden

Beautiful Machines

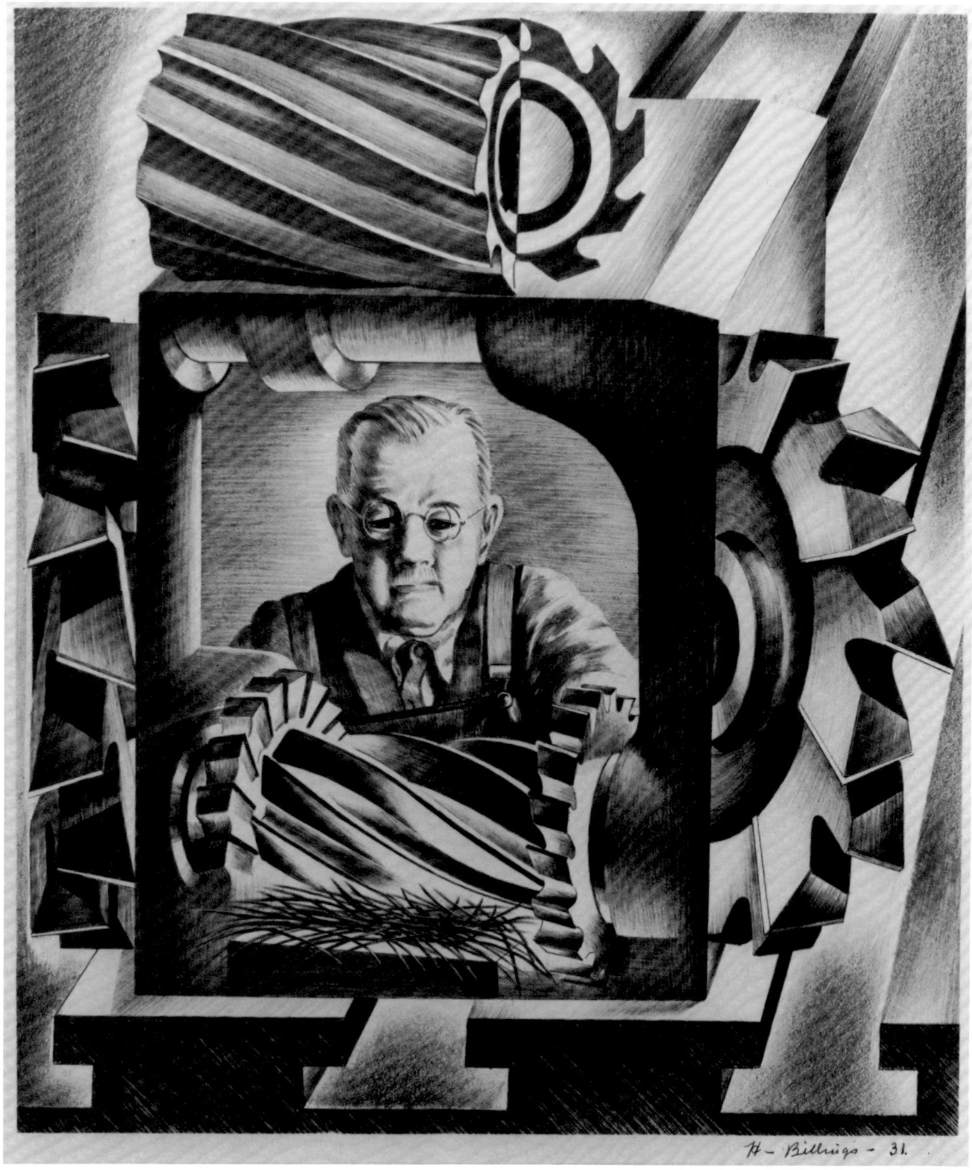

plate 25

Henry Billings

Machines and Men from the series 'Machines and Men,' 1931

lithograph, sheet: 18½ x 16⅞", image: 15¾ x 13¾"

Collection of Whitney Museum of American Art, New York; Purchase, with funds from the Print Purchase Fund 79.63

plate 26

Ralph Steiner

Portrait of Louis Lozowick, c. 1930

gelatin silver, $9\frac{9}{16}$ x $7\frac{5}{8}$″

Lent by The Metropolitan Museum of Art, John B. Turner Fund, 1975

plate 27

Paul Strand

Gear, New York, c. 1920

gelatin silver print, printed from the original negative by Richard Benson, 8 x 10″

Aperture Foundation, Inc., Paul Strand Archive

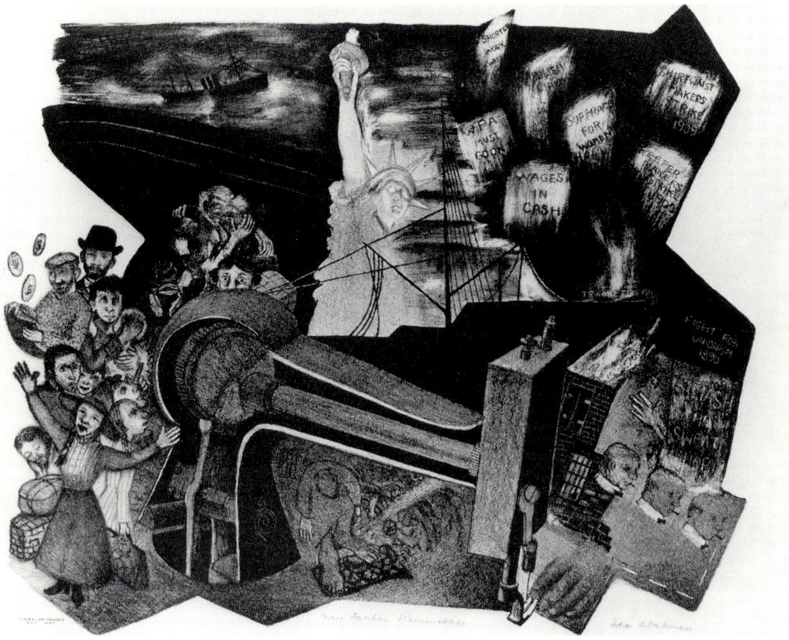

plate 28
Ida Abelman
My Father Reminisces, 1937
lithograph on paper, 15 x 18¼″
National Museum of American Art, Smithsonian Institution, Transfer from D.C. Public Library

plate 29
Helen Lundeberg
The Mirror: Enigma, 1934
oil on celotex, 30 x 24″
The Nora Eccles Harrison Museum of Art, Utah State University, Logan, Utah (Gift of the Marie Eccles Caine Foundation)

plate 30
Walter Murch
The Calculator, 1949
oil on canvas on board, 20 x 20"
Collection IBM Corporation, Armonk, New York

plate 31
Alfred Stieglitz
A Dirigible, 1910
photogravure, 7 x 7″
Lewis Lehr

plate 32
Alfred Stieglitz
Asphalt Paver, 1913
photogravure, 5½ x 7″
Photofind Gallery, New York

plate 33
Anonymous
Hawkins Mitering Machine, c. 1870s
cast-iron frame, rough wood base, 44 x 24 x 22½″
National Museum of American History, Smithsonian Institution

plate 34
Joseph Maresca
The Fruits of Our Labor, 1987
oil and wax on linen, 60 x 50″
Joseph Maresca/The Barbara Fendrick Gallery, New York

Impossible Machines, Anti-mechanical Constructions, Robots

plate 35
Barbara Kruger
Untitled (Your manias become science), 1981
photograph, 37 x 50″
Collection: Dr. Peter and Eileen Broido, West Chicago
Courtesy: Rhona Hoffman Gallery, Chicago

plate 36
Bruce Charlesworth
Untitled (#25 from the "Fate" series), 1986
cibachrome, 23 x 23"
Thomas Barry Fine Arts, New York

plate 37
Orshi Drozdik
The Black Mirror, 1988
black and white photograph, wood frame, anthracite, 102 x 64½ x 13"
Collection of Vera List, Courtesy Tom Cugliani Gallery, New York

plate 38
Tishan Hsu
Body Scanner #1, 1988
plexiglass, alkyd enamel on wood, 16 x 20″
Timothy Greenfield-Sanders

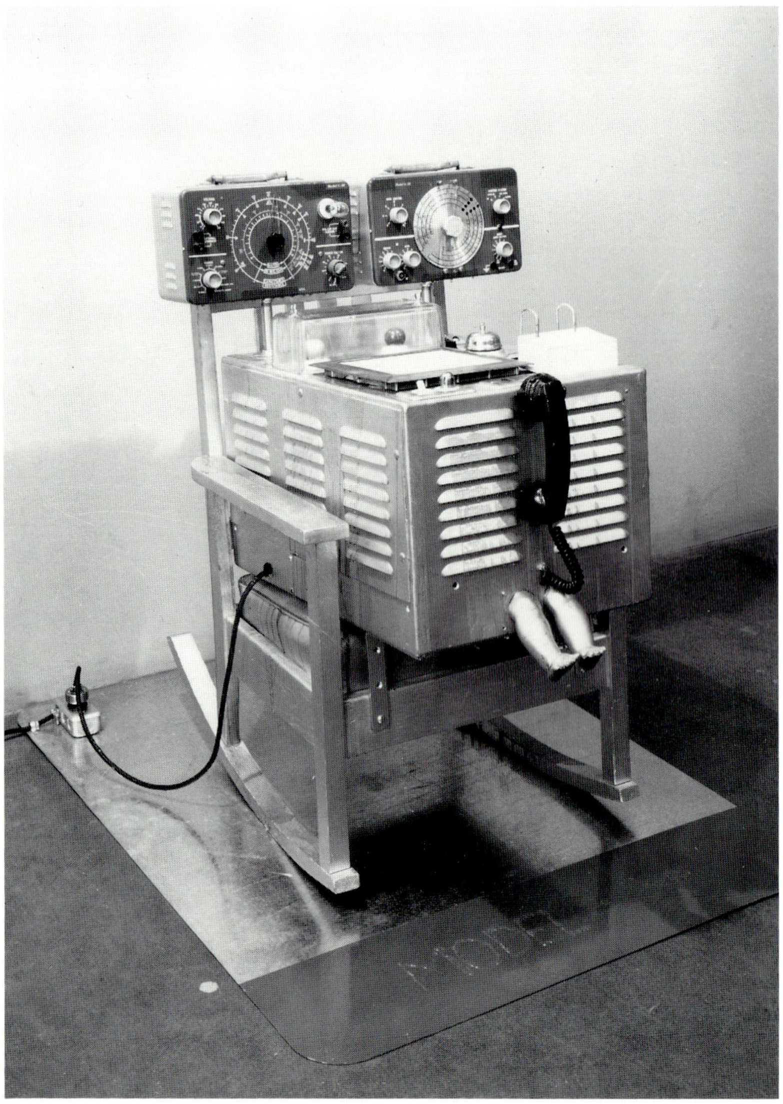

plate 39
Edward Kienholz
The Friendly Grey Computer Star Gauge, #54, 1965
motor-driven assemblage, 40 x 39⅛ x 24½″ on aluminum sheet 48⅛ x 36″

Collection, The Museum of Modern Art, New York. Gift of Jean and Howard Lipman.

plate 40
Chris Macdonald
Work Truck Variation, 1989
oak wood, 89 x 23 x 65"
Courtesy of David Beitzel Gallery, New York

plate 41
Cady Noland
Cart Full of Action, 1986
industrial cart, hubcap, car parts, 40 x 25 x 22"
Courtesy John Gibson Gallery, New York

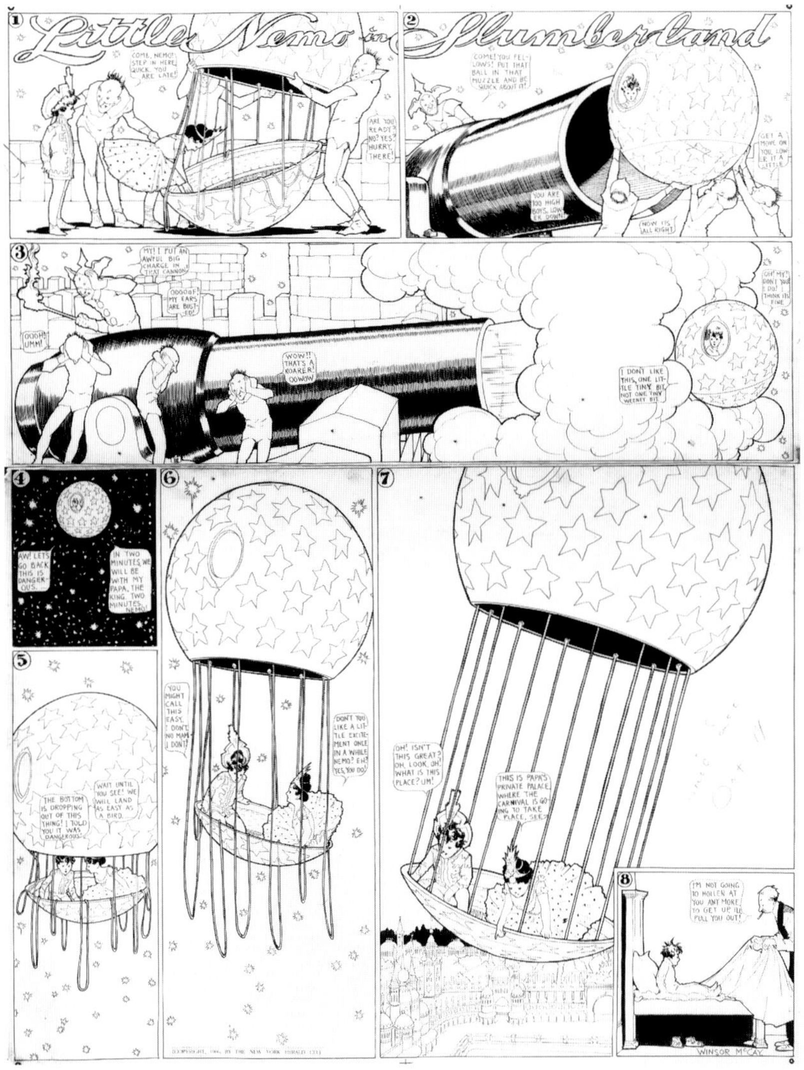

plate 42
Winsor McCay
Little Nemo in Slumberland, October 14, 1906
ink on paper/board, 28½ x 22½"
Ray Winsor Moniz

plate 43
Man Ray
Object Without Title or Date
metal assemblage, 23 x 7¼"
Zabriskie Gallery, New York

plate 44
Hannah Wilke
Hannah Wilke Through The Large Glass, 1976
16 mm film

Philly, 1976
video
Collection of the artist

plate 45
Donald Lipski
Bad Buoy, 1988
ocean mooring buoy, dice, silicone, 5' diameter, 700 lbs.
Courtesy of the artist

plate 46

Claes Oldenburg

Ice Bag–Scale B, 1971

yellow nylon material, fiberglass, and mechanical movement, 40 x 48″, edition of 25

Collaborators: Kenneth Tyler, Jeff Sanders, Lou Faibish, Frank Doose, Roger Merians, Copyright Gemini G.E.L., Los Angeles, California

plate 47
Jasper Johns
Light Bulb, 1960
painted bronze, 2 parts, 4¼ x 6 x 4″
Courtesy Leo Castelli

plate 48
Robert Rauschenberg
Jellyfish Traffic, 1986
3 traffic lights, 40¼ x 31 x 34″, 62 x 40 x 43″, 51 x 29½ x 34″
Courtesy Leo Castelli

plate 49

Lewis Wickes Hine

Junkies Breaking Up Old Looms for Scrap, Paterson, New Jersey, 1937

photograph, 5 x 7"

Photofind Gallery, New York

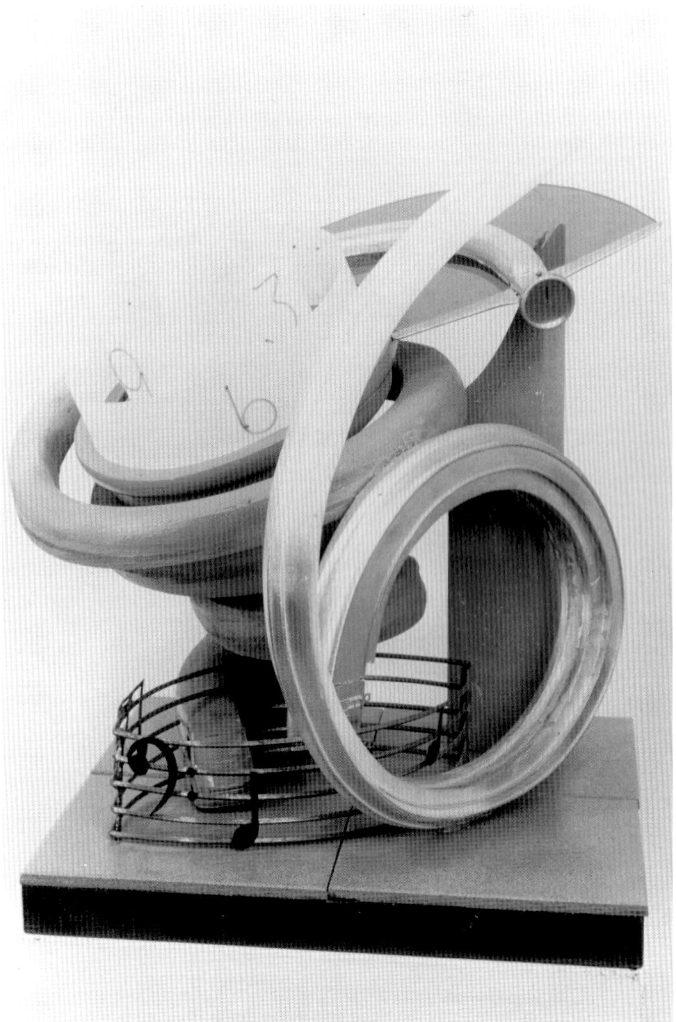

plate 50

Dennis Oppenheim

Collision Wake (Scale Model for CN Plaza, Toronto proposal), 1989

wood, tile, plastic, rubber, model airplane, boat, car, 18 x 17 x 17"

Courtesy Holly Solomon Gallery, New York

plate 51
Alice Aycock
DNA Cutter, 1984
mixed media installation, 85 x 36 x 48"
Courtesy John Weber Gallery, New York

plate 52

Lydia Dona

The Principles of Rupture And The Fluids Of The Limit, 1989

oil, acrylic, and sign paint on canvas, 90 x 72″

Tom Cugliani Gallery, New York

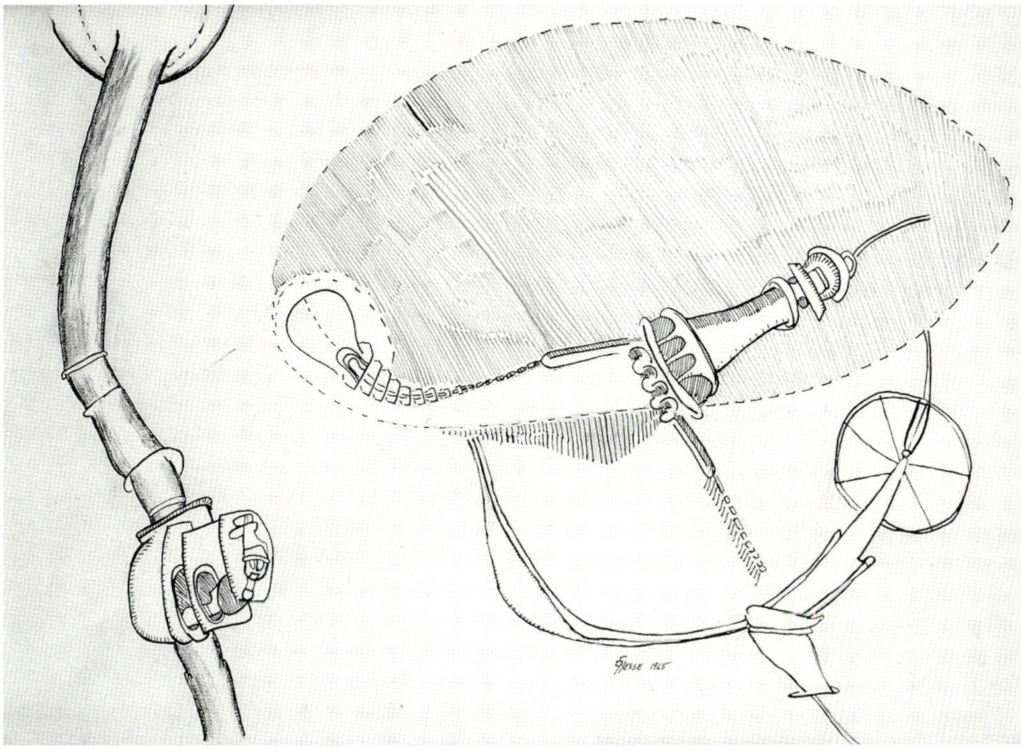

plate 53
Eva Hesse
Untitled, 1965
ink, colored ink on paper, 11½ × 16½″
Robert Miller Gallery, New York

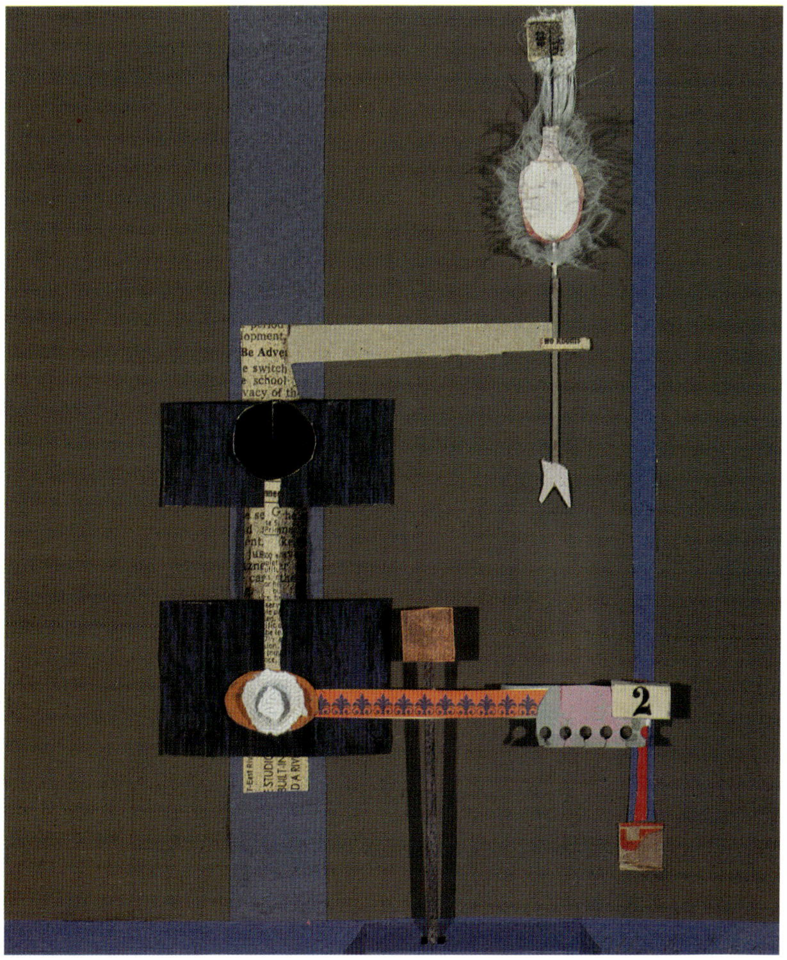

plate 54
Addie Herder
Machine for Luddites, 1983
collage, 13⅞ x 12″
Collection of the artist

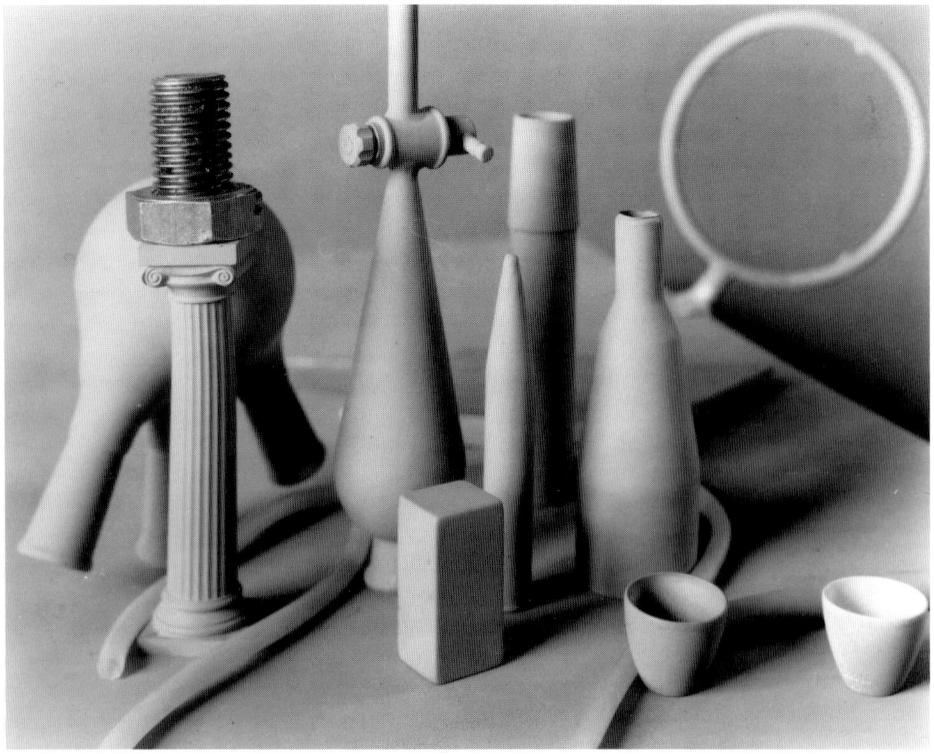

plate 55
Jan Groover
Untitled [Objects with Tube & Circle], 1986
gelatin silver print, 7¾ x 9⅞"
Robert Miller Gallery, New York

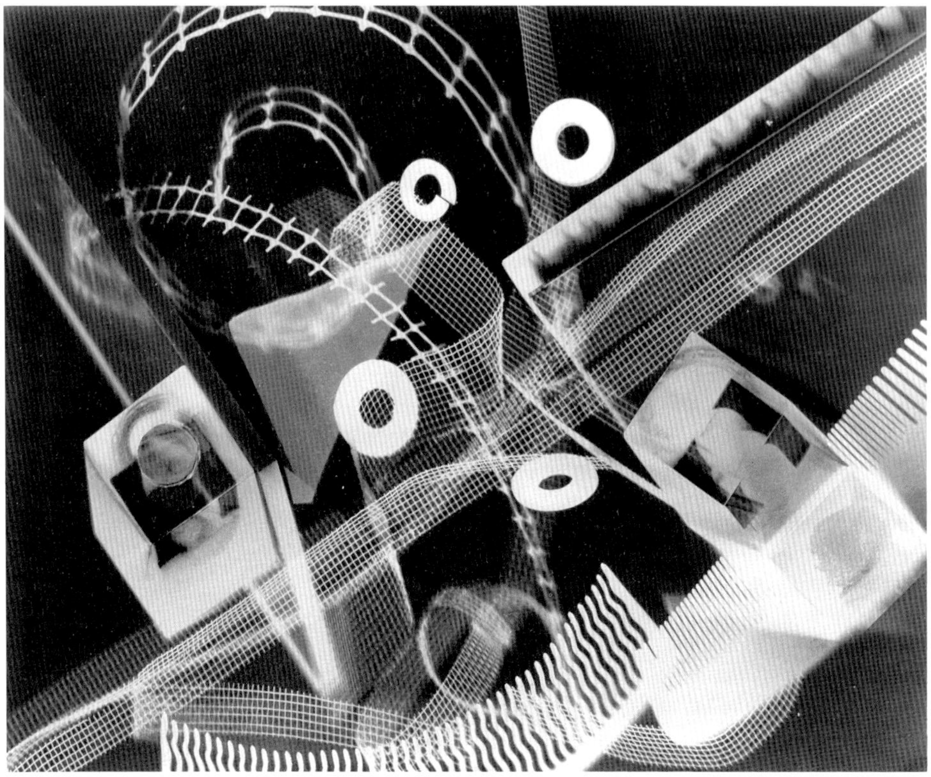

plate 56
Theodore Roszak
Photogram, 1937-39
gelatin silver print, 7⅞ x 9⅞"
Zabriskie Gallery, New York

plate 57
Ron Rocco
The Waterline Project, 1989
steel, fiberglass, wood, glass, organic materials
two sculptures, each approx. 12 x 6 x 6'
Collection of the artist

plate 58
Medrie MacPhee
Press, 1989
oil on canvas, 63 x 90″
Courtesy of Phillipe Daverio Gallery, New York

plate 59
Judith K. Brodsky
The Garbage Bag Mountain, 1989
color lithograph, 47 x 31"
Susan Teller Gallery, New York

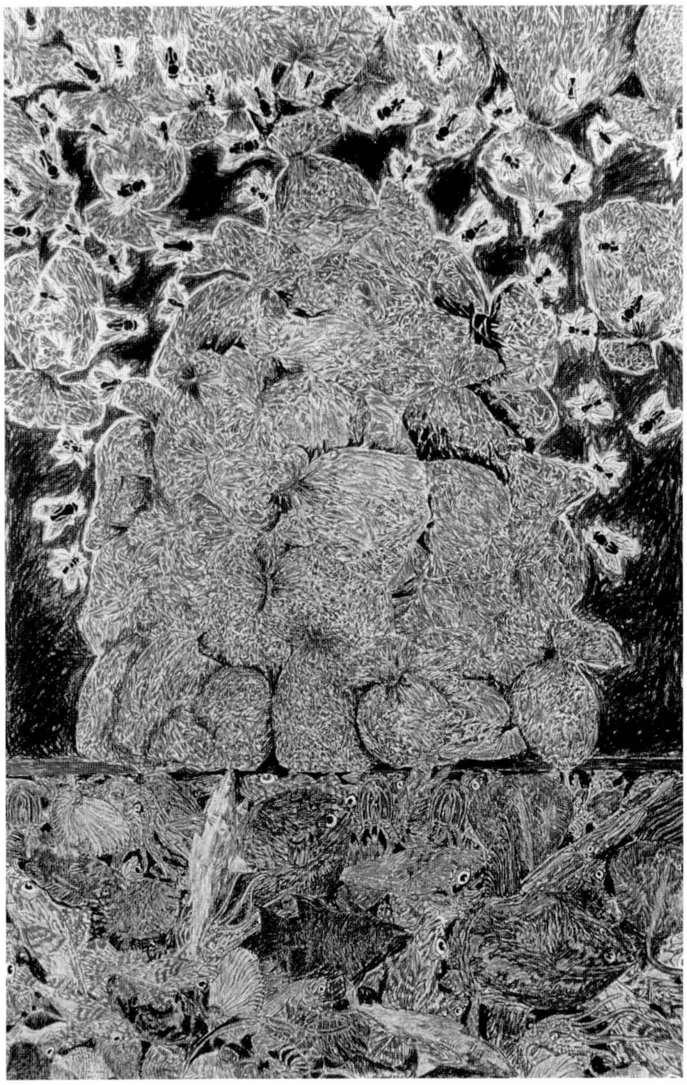

plate 60
Andy Warhol
Evan Picone, Save 20%, 1976-86
four stitched gelatin silver prints
11 x 14" each, overall framed size 31 x 37"
Robert Miller Gallery, New York

The Industrial Landscape and the Ecology

plate 61
Edward Weston
Pipes and Stacks (Armco Steel, Ohio), 1922
gelatin silver print, 9⁷⁄₁₆ x 7⁹⁄₁₆″
San Francisco Museum of Modern Art, Albert M. Bender Collection, Albert M. Bender Bequest Fund, Purchase

plate 62
Larry Gipe
Untitled #5 (Factory Series), 1987
oil on wood panel, 48 x 96″
Courtesy Karl Bornstein Gallery, Santa Monica, California

plate 63
Charles Sheeler
Conversation—Sky and Earth, 1940
oil on canvas, 28 x 23″
The Regis Collection, Minneapolis, Minnesota

plate 64
Denny R. Moers
Figment from a Landscape, 1989
photographic monoprint, 20 x 24″
Courtesy Carlo Lamagna Gallery, New York

plate 65
Ray Mortenson
Newark, Harrison, Amtrak Bridge from South Bank Passaic River, 1982
photograph, 20 x 24″
Ray Mortenson

plate 66
Ralston Crawford
Untitled (Boiler Synthesis), c. 1940
vintage gelatin silver print, 10 x 8″
Robert Miller Gallery, New York

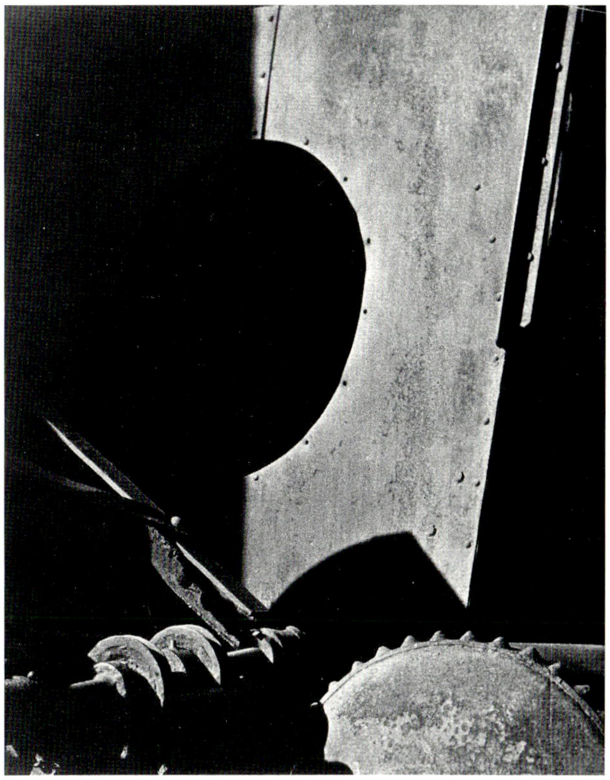

plate 67
Ralston Crawford
Untitled, 1942
oil on board, 8½ x 10½″
Hirschl & Adler Galleries, New York

plate 69

Walker Evans

View of Easton, Pennsylvania on the Lehigh River, November 1935

photograph, 7¾ x 9⅞"

The New York Public Library, The Branch Libraries, Mid-Manhattan Library Picture Collection

plate 68

Ellen Levy

Industry, 1989

oil on panels, *Tidal* 17 x 13″, *Thermal* 12 x 16″, *Potential* 13 x 17″

Collection of the artist

plate 70

Alexander H. Wyant

Scene on the Upper Susquehanna, 1868

oil on canvas, 12 x 26″

From the private collection of Mr. and Mrs. John Ong

plate 71
Margaret Bourke-White
Smokestacks, Otis Steel Company, Cleveland, 1928
photograph, 12⁵⁄₁₆ x 7⁵⁄₈"
George Arents Research Library for Special Collections at Syracuse University

plate 72
Margaret Bourke-White
Slag Piles, Otis Steel, 1928
photograph, 10⁷⁄₈ x 13⁷⁄₈"
George Arents Research Library for Special Collections at Syracuse University

plate 73

Tony Oursler

Psychomimetiscape II, 1988

table top construction with two monitors, 3½' high

Courtesy of the artist and Diane Brown Gallery, New York

Factory Life and the Worker

plate 74
Harold Costain
Meat Packing, 1932
photograph, 13¾ x 10¾″
Photofind Gallery, New York

plate 75
Phillip Evergood
Through the Mill, 1940
oil on canvas, 36 x 52″
Collection of Whitney Museum of American Art.
Purchase 41.24

plate 76
Lewis Wickes Hine
Heart of the Turbine, 1930
gelatin silver print, 4¾ x 6⅞″
Collection of the International Museum of Photography at George Eastman House

plate 77
Waylande DeSantis Gregory
Farmer and Factory Worker, 1939
terra cotta, 56 x 37 x 37″ (21½″ base)
Private Collection

plate 78
Louis Lozowick
Dusk, 1932
oil on canvas, 24 x 16"
Courtesy of Sid Deutsch Gallery, New York

plate 79
Charles Sheeler
Industrial Architecture, 1931
conté crayon on paper, 9½ x 7½"
James Maroney, New York

plate 80
Abraham Joel Tobias
Machine Design, 1934
oil, 48 x 67 x 2⅔"
Collection of the artist Abraham Joel Tobias

plate 81
O. Louis Guglielmi
Relief Blues, 1938
tempera on fiberboard, 24 x 29⅞″
National Museum of American Art, Smithsonian Institution, Transfer from The Museum of Modern Art

plate 82
Lewis Wickes Hine
West Virginia: Small Boys Worked All Night in the Glass Factories, June 1911
photograph, 4½ x 6⅝″
From the collection of Naomi and Walter Rosenblum

The Industrialized Cityscape: Skyscrapers and Bridges

plate 83
Alvin Langdon Coburn
New York City, 1910
photogravure, 6¾ x 6¼"
Photofind Gallery, New York

plate 84
H.C. Torrence
Pittsburgh, c. 1920
photograph, 9 x 11½"
Photofind Gallery, New York

plate 85
Anton Breuhl
Empire State Building, c. 1938
gelatin silver print, 10 x 8″
Robert Miller Gallery, New York

plate 86
Abraham Walkowitz
Cityscape, c. 1910
pencil on paper, 13¾ x 8½″
Zabriskie Gallery, New York

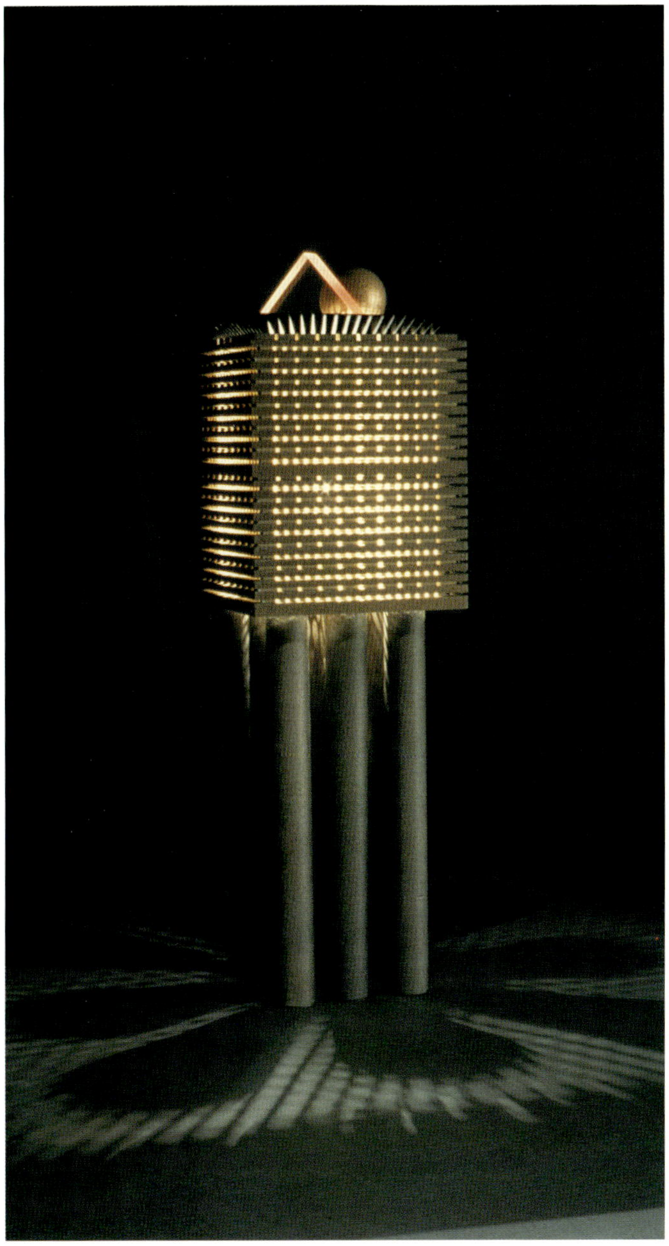

plate 87
James Evanson
High Beam, 1984
plexiglass, lucite; 24 x 8 x 8″
Collection of the artist

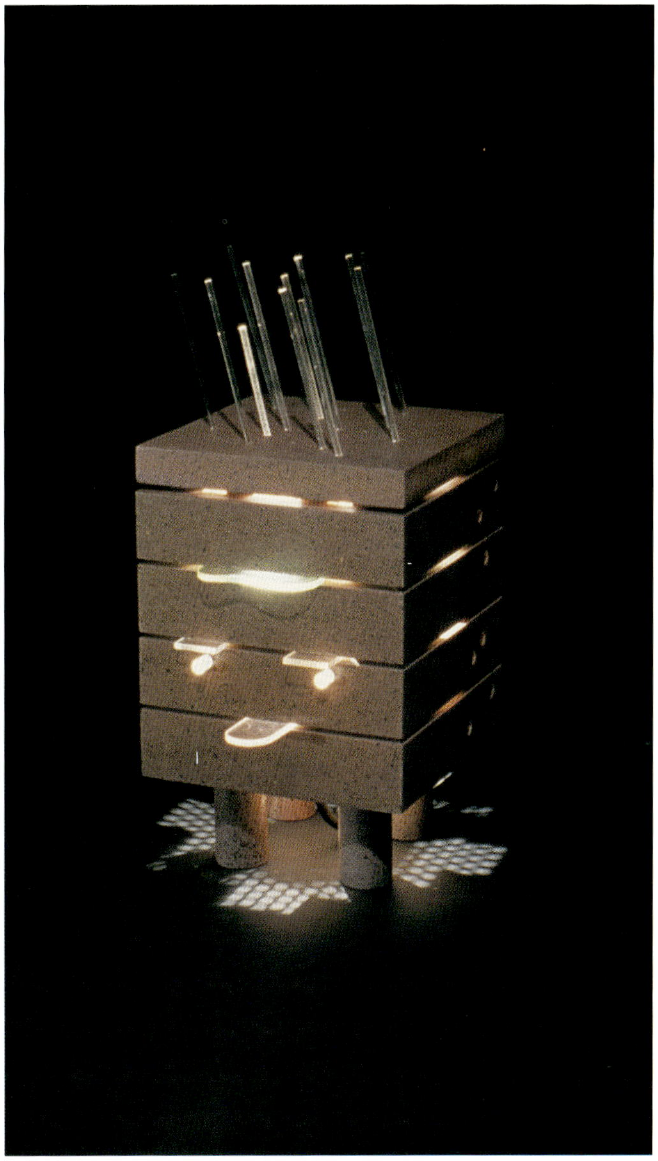

plate 88
James Evanson
Flat Top, 1984
plexiglass, lucite, 8½ x 6 x 6″
Collection of the artist

plate 89
Peter Stackpole
Abstraction From Under 40 Tons, 1934
gelatin silver print, 6⅛ x 9⅛″
The Oakland Museum, California, Gift of Dr. and Mrs. Dudley P. Bell

plate 90
Yvonne Jacquette
Reflected Facade (White Hall Street) II, 1987
oil on three canvas panels
each panel 62¼ x 46″, overall 62¼ x 138″
Courtesy Brooke Alexander, New York

plate 91
Karl Struss
Brooklyn Bridge, c. 1912
photograph (1978 palladium print by Herb Quick), 3¾ x 4¾″
Photofind Gallery, New York

Machinery Domesticated: The Machine Enters the Home

plate 92
Kem Weber
Skyscraper Vanity and Chair, 1929
wood, glass, mirror
vanity 81H x 41W x 17"D, chair 18H x 25W x 17½"D
George Waterman III

plate 93
Kem Weber
Airline Chair, 1934-35
carved birch with molded seat and back, 31 x 25 x 28"
Alan Moss Studios, New York

plate 94
Style of John Henry Belter
Upholstered Rococo Revival Side Chair, 1858-1868
rosewood, ash, 38 x 19 x 17¾"
Lent by The Metropolitan Museum of Art, Gift of Mrs. Charles Reginald Leonard, 1957

plate 95

Paul Ludick

Arms Chair, 1986

mixed media, 36 x 43 x 37″

Loretta Michaelcheck/Art et Industrie, New York

plate 96

Jessica Stockholder

Wood, Lamps, Paint, Sheetrock, Small Coffee Table, 1987

48″ high

Ruth Kaufmann

plate 97

R.M. Fischer

Desk Set, 1987-88

aluminum, cast iron, steel, brass and electric lights, 59⅞ x 48⅛ x 32⅞"

Collection of Whitney Museum of American Art. Purchase, with funds from Mrs. Percy Uris Purchase Fund 88.33

plate 98

William Stone

Fall Line, 1989

wood, brass, water, 53½ x 38 x 32"

The Boardroom/Edelston Collection, Courtesy Tom Cugliani Gallery

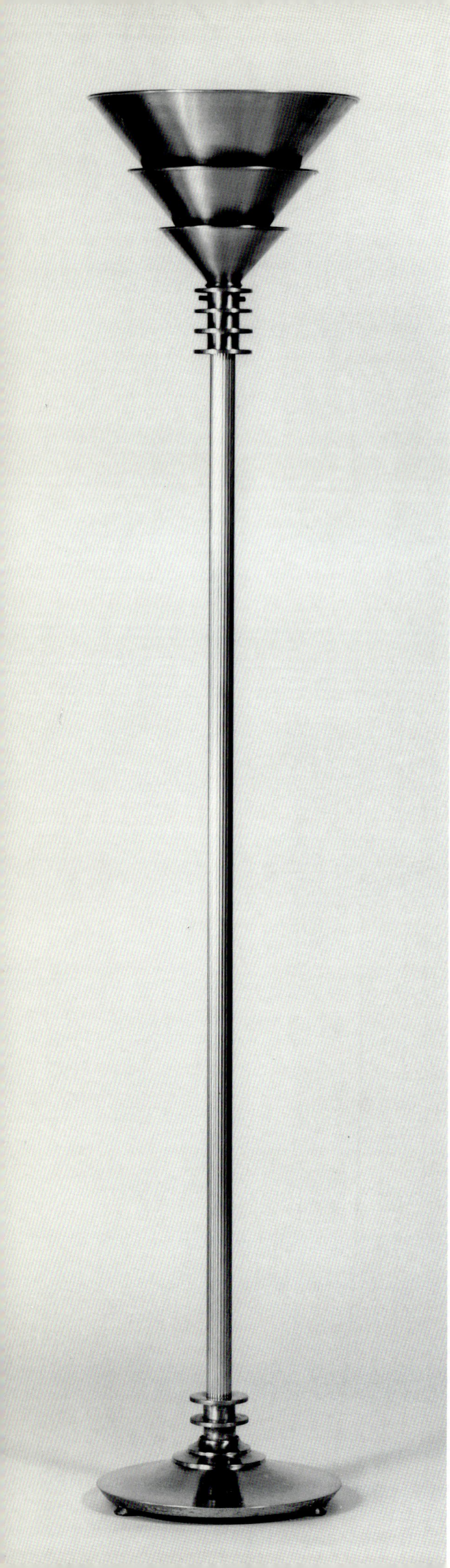

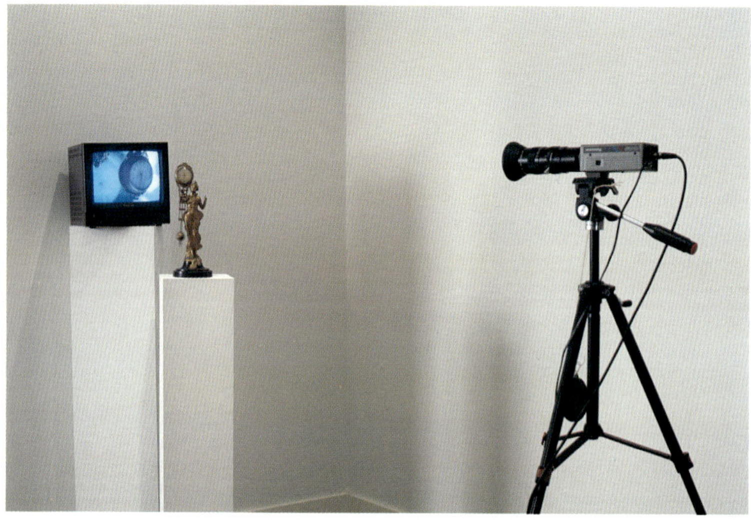

plate 99

Nam June Paik

Art Deco Clock, 1989

1 color monitor, 1 video camera, 1 antique clock,
59 x 100 x 36″

Courtesy Holly Solomon Gallery, New York

plate 100

Walter Von Nessen

Standing Floor Lamp, c. 1928

brushed chrome over brass, iron base

67″ high, 13½″ diam. at top, 12¾″ diam. at bottom

Yale University Art Gallery, Enoch Vine Stoddard,
B.A. 1905, and Marie-Antoinette Slade Funds

plate 101
Forrest Myers
No Evil Outdoor Bench, 1985
one sheet of corten steel plate, 19 x 81 x 16″
Forrest Myers/Art et Industrie, New York

plate 102
Forrest Myers
Love Bench Garden Bench, 1987
folded from one piece of stainless steel,
17 x 80 x 28″
Forrest Myers/Art et Industrie, New York

Checklist of the Exhibition

height precedes width precedes depth.

Ida Abelman
My Father Reminisces, 1937
lithograph on paper, 15 x 18¼″
National Museum of American Art, Smithsonian Institution, Transfer from D.C. Public Library
plate 28

Anonymous
Hawkins Mitering Machine, c. 1870s
cast-iron frame, rough wood base, 44 x 24 x 22½″
National Museum of American History, Smithsonian Institution
plate 33

John Atherton
Christmas Eve, 1941
oil on canvas, 30¼ x 35″
Collection, The Museum of Modern Art, New York. Purchase.
plate 23

Alice Aycock
DNA Cutter, 1984
mixed media installation, 85 x 36 x 48″
Courtesy John Weber Gallery, New York
plate 51

John Baeder
Maple Diner, 1973
oil on canvas, 42 x 66″
Dick Belger Collection, Courtesy Morgan Gallery, Kansas City
plate 17

Style of John Henry Belter
Upholstered Rococo Revival Side Chair, 1858-1868
rosewood, ash, 38 x 19 x 17¾″
Lent by The Metropolitan Museum of Art, Gift of Mrs. Charles Reginald Leonard, 1957
plate 94

Henry Billings
Machines and Men from the series 'Machines and Men,' 1931
lithograph, sheet: 18½ x 16⅞″, image: 15¾ x 13¾″
Collection of Whitney Museum of American Art, New York; Purchase, with funds from the Print Purchase Fund 79.63
plate 25

Margaret Bourke-White
Smokestacks, Otis Steel Company, Cleveland, 1928
photograph, 12⁵⁄₁₆ x 7⅝″
George Arents Research Library for Special Collections at Syracuse University
plate 71

Slag Piles, Otis Steel, 1928
photograph, 10⅞ x 13⅞″
George Arents Research Library for Special Collections at Syracuse University
plate 72

La Salle, c. 1935
photograph, 13⁵⁄₁₆ x 9⅜″
George Arents Research Library for Special Collections at Syracuse University
plate 18

Anton Breuhl
Empire State Building, c. 1938
gelatin silver print, 10 x 8″
Robert Miller Gallery, New York
plate 85

Judith K. Brodsky
The Garbage Bag Mountain, 1989
color lithograph, 47 x 31″
Susan Teller Gallery, New York
plate 59

Bruce Charlesworth
Untitled (#25 from the "Fate" series), 1986
cibachrome, 23 x 23″
Thomas Barry Fine Arts, New York
plate 36

Alvin Langdon Coburn
New York City, 1910
photogravure, 6¾ x 6¼"
Photofind Gallery, New York
plate 83

Harold Costain
Meat Packing, 1932
photograph, 13¾ x 10¾"
Photofind Gallery, New York
plate 74

Ralston Crawford
Untitled (Boiler Synthesis), c. 1940
vintage gelatin silver print, 10 x 8"
Robert Miller Gallery, New York
plate 66

Untitled, 1942
oil on board, 8½ x 10½"
Hirschl & Adler Galleries, New York
plate 67

Lydia Dona
The Principles of Rupture And The Fluids Of The Limit, 1989
oil, acrylic, and sign paint on canvas, 90 x 72"
Tom Cugliani Gallery, New York
plate 52

Orshi Drozdik
The Black Mirror, 1988
black and white photograph, wood frame, anthracite, 102 x 64½ x 13"
Collection of Vera List, Courtesy Tom Cugliani Gallery, New York
plate 37

Thomas Eakins
Models of Horses for "The Fairman Rogers Four-In-Hand," 1879 (cast 1946)
bronze
Left Leader Horse, length 12½"
Right Leader Horse, length 11⅞"
Left Wheeler Horse, length 11¾"
Right Wheeler Horse, length $11\frac{15}{16}$"
Phildelphia Museum of Art. Bequest of Mr. and Mrs. William M. Elkins
plate 2

William England
Niagara Suspension Bridge, 1859
albumen-silver print, 9½ x 11½"
Collection, The Museum of Modern Art, New York. Purchase.
plate 12

Walker Evans
View of Easton, Pennsylvania on the Lehigh River, November 1935
photograph, 7¾ x 9⅞"
The New York Public Library, The Branch Libraries, Mid-Manhattan Library Picture Collection
plate 69

James Evanson
High Beam, 1984
plexiglass, lucite; 24 x 8 x 8"
Collection of the artist
plate 87

Flat Top, 1984
plexiglass, lucite, 8½ x 6 x 6"
Collection of the artist
plate 88

Phillip Evergood
Through the Mill, 1940
oil on canvas, 36 x 52"
Collection of Whitney Museum of American Art. Purchase 41.24
plate 75

John Fekner
Industrial Fossil, 1978, 1990
Reenacted site-specific installation originally spray painted on truck—Maspeth, N.Y.
Courtesy Bjorn Olsson Gallery, Stockholm, Sweden
plate 24

R.M. Fischer
Desk Set, 1987-88
aluminum, cast iron, steel, brass and electric lights, 59⅞ x 48⅛ x 32⅞"
Collection of Whitney Museum of American Art. Purchase, with funds from Mrs. Percy Uris Purchase Fund 88.33
plate 97

Larry Gipe
Untitled #5 (Factory Series), 1987
oil on wood panel, 48 x 96"
Courtesy Karl Bornstein Gallery, Santa Monica, California
plate 62

Waylande DeSantis Gregory
Farmer and Factory Worker, 1939
terra cotta, 56 x 37 x 37" (21½" base)
Private Collection
plate 77

Jan Groover
Untitled [Objects with Tube & Circle], 1986
gelatin silver print, 7¾ x 9⅞"
Robert Miller Gallery, New York
plate 55

O. Louis Guglielmi
Relief Blues, 1938
tempera on fiberboard, 24 x 29⅞"
National Museum of American Art, Smithsonian Institution, Transfer from The Museum of Modern Art
plate 81

Hananiah Harari
Smiles, 1940
oil on canvas, 10 x 11"
Richard York Gallery, New York
plate 11

Addie Herder
Machine for Luddites, 1983
collage, 13⅞ x 12"
Collection of the artist
plate 54

Eva Hesse
Untitled, 1965
ink, colored ink on paper, 11½ × 16½"
Robert Miller Gallery, New York
plate 53

Lewis Wickes Hine
West Virginia: Small Boys Worked All Night in the Glass Factories, June 1911
photograph, 4½ x 6⅝"
From the collection of Naomi and Walter Rosenblum
plate 82

Heart of the Turbine, 1930
gelatin silver print, 4¾ x 6⅞"
Collection of the International Museum of Photography at George Eastman House
plate 76

Junkies Breaking Up Old Looms for Scrap, Paterson, New Jersey, 1937
photograph, 5 x 7"
Photofind Gallery, New York
plate 49

Winslow Homer
The Army of the Potomac, Rebel Sharpshooter on Picket Duty, 1862
wood engraving, 9⅛ x 13¾"
Private Collection
plate 3

Tishan Hsu
Body Scanner #1, 1988
plexiglass, alkyd enamel on wood, 16 x 20"
Timothy Greenfield-Sanders
plate 38

George Inness
Delaware Water Gap, 1857
oil on canvas, 32 x 52¼″
The Monclair Art Museum, Gift of Mrs. F.G. Herman Fayen
plate 14

Yvonne Jacquette
Reflected Facade (White Hall Street) II, 1987
oil on three canvas panels
each panel 62¼ x 46″; overall 62¼ x 138″
Courtesy Brooke Alexander, New York
plate 90

Jasper Johns
Light Bulb, 1960
painted bronze, 2 parts, 4¼ x 6 x 4″
Courtesy Leo Castelli
plate 47

David Kapp
Oncoming Car X, 1989
oil on linen, 24 x 48″
Private collection, Courtesy David Beitzel Gallery, New York
plate 20

Edward Kienholz
The Friendly Grey Computer Star Gauge, #54, 1965
motor-driven assemblage, 40 x 39⅛ x 24½″ on aluminum sheet 48⅛ x 36″
Collection, The Museum of Modern Art, New York. Gift of Jean and Howard Lipman.
plate 39

Barbara Kruger
Untitled (Your manias become science), 1981
photograph, 37 x 50″
Collection: Dr. Peter and Eileen Broido, West Chicago
Courtesy: Rhona Hoffman Gallery, Chicago
plate 35

Dorothea Lange
Next Time Try the Train, Relax, c. 1935
photograph, 10¾ x 13¾″
Photofind Gallery, New York
plate 13

Plantation Owner and His Field Hands, Mississippi Delta, 1936
photograph, 10 x 8″
Private Collection
plate 21

Ellen Levy
Industry, 1989
oil on panels, *Tidal* 17 x 13″, *Thermal* 12 x 16″, *Potential* 13 x 17″
Collection of the artist
plate 68

Donald Lipski
Bad Buoy, 1988
ocean mooring buoy, dice, silicone, 5′ diameter, 700 lbs.
Courtesy of the artist
plate 45

Louis Lozowick
Dusk, 1932
oil on canvas, 24 x 16″
Courtesy of Sid Deutsch Gallery, New York
plate 78

Paul Ludick
Arms Chair, 1986
mixed media, 36 x 43 x 37″
Loretta Michaelcheck/Art et Industrie, New York
plate 95

Helen Lundeberg
The Mirror: Enigma, 1934
oil on celotex, 30 x 24″
The Nora Eccles Harrison Museum of Art, Utah State University, Logan, Utah (Gift of the Marie Eccles Caine Foundation)
plate 29

Chris Macdonald
Work Truck Variation, 1989
oak wood, 89 x 23 x 65″
Courtesy of David Beitzel Gallery, New York
plate 40

Medrie MacPhee
Press, 1989
oil on canvas, 63 x 90″
Courtesy of Phillipe Daverio Gallery, New York
plate 58

Joseph Maresca
The Fruits of Our Labor, 1987
oil and wax on linen, 60 x 50″
Joseph Maresca/The Barbara Fendrick Gallery, New York
plate 34

Winsor McCay
Little Nemo in Slumberland, October 14, 1906
ink on paper/board, 28½ x 22½″
Ray Winsor Moniz
plate 42

Denny R. Moers
Figment from a Landscape, 1989
photographic monoprint, 20 x 24″
Courtesy Carlo Lamagna Gallery, New York
plate 64

Barbara Morgan
Valerie Bettis Desperate Heart—II, 1944
photograph, 8½ x 6¹¹⁄₁₆″
Barbara Morgan
plate 10

Wild Bee Honeycomb/Skyscraper, 1973
photomontage, 9½ x 6⅜″
Barbara Morgan
frontispiece

Ray Mortenson
Newark, Harrison, Amtrak Bridge from South Bank Passaic River, 1982
photograph, 20 x 24″
Ray Mortenson
plate 65

Walter Murch
The Calculator, 1949
oil on canvas on board, 20 x 20″
Collection IBM Corporation, Armonk, New York
plate 30

Eadweard Muybridge
"Head Spring, a flying pigeon interfering," from *Animal Locomotion*, vol. 5, plate 365, 1887
photograph, Image size: 8¼ x 14½″, Sheet size: 18¼ x 23″
The Miriam and Ira D. Wallach Division of Art, Prints and Photographs, The New York Public Library, Astor, Lenox and Tilden Foundations
plate 1

Forrest Myers
No Evil Outdoor Bench, 1985
one sheet of corten steel plate, 19 x 81 x 16″
Forrest Myers/Art et Industrie, New York
plate 101

Love Bench Garden Bench, 1987
folded from one piece of stainless steel, 17 x 80 x 28″
Forrest Myers/Art et Industrie, New York
plate 102

Cady Noland
Cart Full of Action, 1986
industrial cart, hubcap, car parts, 40 x 25 x 22″
Courtesy John Gibson Gallery, New York
plate 41

Claes Oldenburg
Ice Bag—Scale B, 1971
yellow nylon material, fiberglass, and mechanical movement, 40 x 48″, edition of 25
Collaborators: Kenneth Tyler, Jeff Sanders, Lou Faibish, Frank Doose, Roger Merians, Copyright Gemini G.E.L., Los Angeles, California
plate 46

Dennis Oppenheim
Collision Wake (Scale Model for CN Plaza, Toronto proposal), 1989
wood, tile, plastic, rubber, model airplane, boat, car, 18 x 17 x 17″
Courtesy Holly Solomon Gallery, New York
plate 50

Timothy O'Sullivan
Destruction of a Railroad Bridge, (North Anna River, May 25, 1864)
albumen print, stereopticon view, 3 x 7″
Library of Congress
plate 5

Devil's Slide Bridge, c. 1870
albumen print, $7\frac{15}{16}$ x $10\frac{5}{8}$″
Library of Congress
plate 4

Devil's Slide, Weber Canyon, Utah, c.1870
albumen print, $10\frac{5}{8}$ x $7\frac{7}{8}$″
Library of Congress
plate 6

Devil's Slide, Weber Canyon, Utah, c. 1870
albumen print, $7\frac{7}{8}$ x $10\frac{5}{8}$″
Library of Congress
plate 7

Devil's Slide, Weber Canyon, Utah, c. 1870
albumen print, $7\frac{14}{16}$ x $10\frac{5}{8}$″
Library of Congress
plate 8

Tony Oursler
Psychomimetiscape II, 1988
table top construction with two monitors, $3\frac{1}{2}$′ high
Courtesy of the artist and Diane Brown Gallery, New York
plate 73

Nam June Paik
Art Deco Clock, 1989
1 color monitor, 1 video camera, 1 antique clock, 59 x 100 x 36″
Courtesy Holly Solomon Gallery, New York
plate 99

John Cage, 1990
mixed media sculpture with video, approx. 7′
Collection of Carl and Elizabeth Solway
back cover

Robert Rauschenberg
Jellyfish Traffic, 1986
3 traffic lights, $40\frac{1}{4}$ x 31 x 34″, 62 x 40 x 43″, 51 x $29\frac{1}{2}$ x 34″
Courtesy Leo Castelli
plate 48

Man Ray
Object Without Title or Date
metal assemblage, 23 x $7\frac{1}{4}$″
Zabriskie Gallery, New York
plate 43

Ron Rocco
The Waterline Project, 1989
steel, fiberglass, wood, glass, organic materials
two sculptures, each approx. 12 x 6 x 6′
Collection of the artist
plate 57

John Rogers
The Photographer and the Sitter, 1878
painted plaster
The Photographer: $18\frac{1}{2}$ x $9\frac{1}{2}$″, The Sitter: 17 x $8\frac{1}{2}$″
Courtesy of The Historical Society of the Town of Greenwich
plate 9

James Rosenquist
Untitled (Broome Street Truck), 1963
oil on canvas, 72 x 72″
Collection of Whitney Museum of American Art. Purchase 64.20
plate 22

Theodore Roszak
Man at Machine, 1937
oil on canvas, 24 x $40\frac{1}{8}$″
The Regis Collection, Minneapolis, Minnesota
cover

Photogram, 1937-39
gelatin silver print, $7\frac{7}{8}$ x $9\frac{7}{8}$″
Zabriskie Gallery, New York
plate 56

Mrs. Saunders
Train Quilt Top, c. 1884
pieced and appliqued cotton, $101\frac{1}{4}$ x $90\frac{1}{4}$″
Collection of Mississippi State Historical Museum
plate 15

Charles Sheeler
Industrial Architecture, 1931
conté crayon on paper, $9\frac{1}{2}$ x $7\frac{1}{2}$″
James Maroney, New York
plate 79

Conversation—Sky and Earth, 1940
oil on canvas, 28 x 23″
The Regis Collection, Minneapolis, Minnesota
plate 63

Peter Stackpole
Abstraction From Under 40 Tons, 1934
gelatin silver print, $6\frac{1}{8}$ x $9\frac{1}{8}$″
The Oakland Museum, California, Gift of Dr. and Mrs. Dudley P. Bell
plate 89

Ralph Steiner
Portrait of Louis Lozowick, c. 1930
gelatin silver, $9\frac{9}{16}$ x $7\frac{5}{8}$″
Lent by The Metropolitan Museum of Art, John B. Turner Fund, 1975
plate 26

Alfred Stieglitz
A Dirigible, 1910
photogravure, 7 x 7″
Lewis Lehr
plate 31

Asphalt Paver, 1913
photogravure, $5\frac{1}{2}$ x 7″
Photofind Gallery, New York
plate 32

Jessica Stockholder
Wood, Lamps, Paint, Sheetrock, Small Coffee Table, 1987
48″ high
Ruth Kaufmann
plate 96

William Stone
Fall Line, 1989
wood, brass, water, $53\frac{1}{2}$ x 38 x 32″
The Boardroom/Edelston Collection, Courtesy Tom Cugliani Gallery
plate 98

Paul Strand
Gear, New York, c. 1920
gelatin silver print, printed from the original negative by Richard Benson, 8 x 10″
Aperture Foundation, Inc., Paul Strand Archive
plate 27

Karl Struss
Brooklyn Bridge, c. 1912
photograph (1978 palladium print by Herb Quick), $3\frac{3}{4}$ x $4\frac{3}{4}$″
Photofind Gallery, New York
plate 91

Abraham Joel Tobias
Machine Design, 1934
oil, 48 x 67 x $2\frac{2}{3}$″
Collection of the artist Abraham Joel Tobias
plate 80

George Tooker
The Subway, 1950
egg tempera on composition board, 18 x 36"
Collection of Whitney Museum of American Art. Purchase, with funds from the Juliana Force Purchase Award. 50.23
plate 16

H.C. Torrence
Pittsburgh, c. 1920
photograph, 9 x 11½"
Photofind Gallery, New York
plate 84

Walter Von Nessen
Standing Floor Lamp, c. 1928
brushed chrome over brass, iron base
67" high, 13½" diam. at top, 12¾" diam. at bottom
Yale University Art Gallery, Enoch Vine Stoddard, B.A. 1905, and Marie-Antoinette Slade Funds
plate 100

Abraham Walkowitz
Cityscape, c. 1910
pencil on paper, 13¾ x 8½"
Zabriskie Gallery, New York
plate 86

Andy Warhol
Evan Picone, Save 20%, 1976-86
four stitched gelatin silver prints
11 x 14" each, overall framed size 31 x 37"
Robert Miller Gallery, New York
plate 60

Kem Weber
Skyscraper Vanity and Chair, 1929
wood, glass, mirror
vanity 81H x 41W x 17"D, chair 18H x 25W x 17½"D
George Waterman III
plate 92

Airline Chair, 1934-35
carved birch with molded seat and back, 31 x 25 x 28"
Alan Moss Studios, New York
plate 93

Dan Weiner
Ross Sams, Church Furniture Maker, Waco, Texas, 1955
photograph, 9 x 13½"
Private Collection
plate 19

Edward Weston
Pipes and Stacks (Armco Steel, Ohio), 1922
gelatin silver print, 9 7/16 x 7 9/16"
San Francisco Museum of Modern Art, Albert M. Bender Collection, Albert M. Bender Bequest Fund, Purchase
plate 61

Hannah Wilke
Hannah Wilke, Through The Large Glass, 1976
16 mm film

Philly, 1976
video
Collection of the artist
plate 44

Alexander H. Wyant
Scene on the Upper Susquehanna, 1868
oil on canvas, 12 x 26"
From the private collection of Mr. and Mrs. John Ong
plate 70

Nina Yankowitz and Barry Holden
Untitled: Painting with Moving Sphere, 1990
fresco, mixed mediums, 2' x 5' x 27"
Collection of the artists
plate 103

Selected Bibliography

Adams, Henry. *The Education of Henry Adams*. New York: Modern Library, 1931.

Agar, John A. "The New Emphasis in American Industrial Art." *An International Exposition of Art and Industry, Macy's, New York* (May 1928): 64-76.

Anderson, Ross and Barbara Perry. *The Diversions of Keramos: American Clay Sculpture, 1925-1950*. Syracuse, New York: Everson Museum of Art, 1983.

Aycock, Alice. *Alice Aycock: Retrospective of Projects and Ideas, 1972-83*. Stuttgart: Würtembergerkunstverein, 1983.

Baeder, John. *Diners*. New York: Abrams, 1983.

Baker, John. O. *Louis Guglielmi: A Retrospective Exhibition*. New Brunswick, New Jersey: The Rutgers University Art Gallery, 1980.

Banham, Reyner. *Theory and Design in the First Machine Age*. New York: Praeger, 1970.

Barthes, Roland. *Mythologies*. New York: Hill and Wang, 1972.

Baur, John I. H. *Revolution and Tradition in Modern American Art*. New York: Praeger, 1951.

Benthall, Jonathan. *Science and Technology in Art Today*. New York: Praeger, 1972.

Berkson, Bill. "Yvonne Jacquette." *Artforum* 27 (September 1988): 138.

Bletter, Rosemary Haag, *et al. Remembering the Future*. New York: Rizzoli in conjunction with the Queens Museum, 1989.

Bourke White, Margaret. *Portrait of Myself*. New York: Simon and Schuster, 1963.

Breslin, Jimmy. "Chair of the Month." [Paul Ludick] *HG* 160 (August 1988): 25.

Brodsky, Judith K. *Recent Drawings and Lithographs including The Meadowlands Strike Back, A Series on the Industrial Landscape*. New York: Susan Teller Gallery, 1989.

Canemaker, John. *Dreams in Motion: The Art of Winsor McCay*. Katonah, New York: The Katonah Gallery, 1988.

Cavanah, P. J., and James Michic. *Vachel Lindsay*. Granite Falls, Minnesota: Spoon River Poetry Press, 1984.

Celant, Germano. *A Bottle of Notes and Some Voyages: Claes Oldenburg and Coosje Van Bruggen*. Leeds, England: Northern Centre for Contemporary Art and The Henry Moore Centre for the Study of Sculpture, 1988.

Chadwick, Whitney. *Women, Art, and Society*. New York: Thames and Hudson, 1990.

Chase, Stuart. *Men and Machines*. New York: Macmillan, 1929.

Cheney, Sheldon and Martha. *Art and the Machine*. New York: McGraw Hill, 1936.

Colorado State Teachers College versus University of Denver. "The Machine Age—Resolved: That we have more to fear than to hope for from the further development of machines." *The Debater's Annual*, New York: H. W. Wilson, 1930-31.

Corn, Joseph J. and Brian Horrigan. *Yesterday's Tomorrows*. New York: Summit Books, 1984.

Damianovic, Maia. "Paradox View: New Paintings by Lydia Dona." *Arts Magazine* 64 (December 1989): 70-73.

Danly, Susan and Leo Marx, eds. *The Railroad in American Art: Representations of Technological Change*. Cambridge, Massachusetts: The Massachusetts Institute of Technology Press, 1988.

Davies, Karen. *At Home In Manhattan*. New Haven, Connecticut: Yale University Art Gallery, 1983.

D'Harnoncourt, Anne and Kynaston McShine, eds. *Marcel Duchamp*. New York: The Museum of Modern Art, 1973.

Doezema, Marianne. *American Realism and the Industrial Age*. Cleveland: Cleveland Museum of Art in cooperation with Indiana University Press,1980.

Dreishpoon, Douglas. *Theodore Roszak*. New York: Hirschl and Adler Galleries, 1989.

Elliott, George P. *Dorothea Lange*. New York: The Museum of Modern Art, 1966.

Fekner, John. *Queensites: Outdoor Stencil Projects*. New York: Fred Baca, 1979.

Ferrero, Pat, Elaine Hedges, and Julie Silber. *Hearts and Hands: The Influence of Women and Quilts on American Society*. San Francisco: The Quilt Digest Press, 1987.

Fillin-Yeh, Susan. "Charles Sheeler and the Machine Age." Ph. D. diss., The Graduate Center, City University of New York, 1981.

Fletcher, Valerie J. *Dreams and Nightmares: Utopian Visions in Modern Art*. Washington, D.C.: The Smithsonian Institution, 1983.

Frank, Peter. *Lawrence Gipe: Paintings*. Santa Monica, California: Karl Bornstein Gallery, 1987.

Frascella, Lawrence, et al. "Technology and Transformation." *Aperture* 106 (Spring 1987).

Gallatin, A. E. "The Problems of the Museum in Contemporary Art." *Space* 1 (March 1930): 6-7.

Giedion, Siegfried. *Mechanization Takes Command*. New York: W. W. Norton, 1969.

Gilbreth, Lillian. "This Machine Age." *The Survey* 61 (October 1928): 37-38.

Gramsci, Antonio. *Selections from the Prison Notebooks*. edited and translated by Quintin Hoare and Geoffrey Nowell Smith. London: Lawrence and Wishart, 1971.

Gutman, Judith. *Lewis Hine—The W.P.A. Project: The Worker and the Machine 1936-37*. New York: Photofind Gallery, 1989.

Hagen, Charles. "Bruce Charlesworth: The Case of the Missing Hero." *Artforum* 26 (April 1988): 131-35.

Hardison, O. B. *Disappearing Through the Skylight: Culture and Technology in the Twentieth Century.* New York: Viking Penguin, 1989.

Haskell, Barbara. *Blam!: The Explosion of Pop, Minimalism, and Performance 1958-1964.* New York: The Whitney Museum of American Art in association with W. W. Norton, 1984.

———. *Ralston Crawford.* New York: The Whitney Museum of American Art, 1985.

Haviland, Paul. "We Are Living in the Age of the Machine." *291 Magazine* 7-8 (September-October 1915).

Heap, Jane. "The Machine Age Exposition." *The Little Review* Supplement (May 1927).

Heilbrun, Carolyn G. *Writing A Woman's Life.* New York: W. W. Norton, 1988.

Holmes, Mark. *It's Not Over 'Til The Fat Lady Sings. Jessica Stockholder.* Vancouver, British Columbia: Contemporary Art Gallery, 1987.

Hulten, K. G. Pontus. *The machine as seen at the end of the mechanical age.* New York: The Museum of Modern Art, 1968.

Institute of Contemporary Art, Philadelphia. *Machineworks: Vito Acconci, Alice Aycock, Dennis Oppenheim.* Philadelphia: University of Pennsylvania Press, 1981.

James, Henry. *The American Scene.* Bloomington: Indiana University Press, 1968.

Johnson, Phillip and Alfred H. Barr, Jr. *Machine Art.* New York: The Museum of Modern Art, 1934.

Kasson, John. *Civilizing the Machine: Technology and Republican Virtues in America.* New York: Grossman, 1976.

Kaufmann, Rick. *Forrest Myers.* New York: Art et Industrie, 1988.

Klingender, Francis D. *Art and the Industrial Revolution.* London: Noel Carrington, 1947.

Kouwenhoven, John A. *The Arts in Modern American Civilization.* New York: W. W. Norton, 1948.

Krauss, Rosalind. *The Originality of the Avant Garde and Other Modernist Myths.* Cambridge, Massachusetts: The Massachusetts Institute of Technology Press, 1985.

Kuhn, Thomas S. *The Structure of Scientific Revolutions.* Chicago: University of Chicago Press, 1964.

Lawrence, Sidney. "Clean Machines at the Modern." *Art in America* 72 (February 1984): 127-41; 166-68.

Leger, Fernand. "The Esthetics of the Machine: Part I." *The Little Review* 9 (Spring 1923): 45-49.

———. "The Esthetics of the Machine: Part II." *The Little Review* 10 (Fall-Winter 1924): 55-58.

Le Guin, Ursula K. *Dancing at the Edge of the World.* New York: Harper and Row, 1989.

Lewis, Arthur O., Jr. *Of Men and Machines.* New York: E. P. Dutton, 1963.

Lippard, Lucy R. *Pop Art.* New York: Praeger, 1966.

———. *Eva Hesse.* New York: New York University Press, 1976

Lippman, Herbert. "The Machine Age Exposition." *The Arts* 11 (June 1927): 324-26.

Lockridge, Katherine, and Susan Fillin-Yeh. *The Precisionist Painters: Interpretations of a Mechanical Age.* Huntington, New York: The Heckscher Museum, 1978.

Maddox, Kenneth W. *In Search of the Picturesque: Nineteenth Century Images of Industry Along the Hudson River Valley.* Annandale-on-Hudson, New York: Bard College, 1983.

Marx, Leo. *The Machine and the Garden: Technology and the Pastoral Ideal in America.* New York: Oxford University Press, 1964.

Mckinzie, Richard D. *The New Deal for Artists.* Princeton, New Jersey: Princeton Univesity Press, 1973.

Mellor, Anne K. *Mary Shelley: Her Life, Her Fiction, Her Monsters.* New York: Methuen, 1988.

Morgan, Barbara. *Barbara Morgan: Photomontage.* Dobbs Ferry, New York: Morgan and Morgan, 1980.

Mortenson, Ray. *Meadowland: Ray Mortenson.* New York: Lustrum Press, 1983.

Mumford, Lewis. "Modernism for Sale." *American Mercury* 16 (January-April 1929): 455.

———. *The Myth of the Machine.* New York: Harcourt, Brace, Jovanovich, 1970.

———. "Smashed." *The Art Digest* 2 (15 October 1927): 8.

Murdock, Robert M. *Ellen K. Levy: Paintings and Drawings.* Berlin: Galerie Zindel-Grabner, 1990.

Neue Gesellschaft für bildende Kunst. *America: Traum und Depression 1920/40.* Akademie der Kunst: Berlin (West), 1980.

Newhall, Beaumont, editor. *Photography: Essays and Images.* New York: The Museum of Modern Art, 1980.

O'Brien, Harrington. *The Dance of the Machines.* New York: The Macaulay Company, 1929.

Oursler, Tony and Phillip K. Dick. *Psychomimetiscape: A Conversation between Tony Oursler and Philip K. Dick.* Düsseldorf: Delta Galerie, 1989.

Park, Marlene and Gerald E. Markowitz. *New Deal For Art.* Hamilton, New York: The Gallery Association of New York State, 1977.

Phillips, Lisa, with essays by Rosemary Haag Bletter *et al. High Styles: Twentieth-Century American Design.* New York: Whitney Museum of American Art in association with Summit Books, 1985.

Pincus-Witten, Robert. "Kapp's Grant." *David Kapp: Recent Paintings and Drawings.* Manhattan Art: New York, 1985.

Pond, Irving. "Zoning and the Architecture of High Buildings." *Architectural Forum* 35 (October 1921): 131-34.

Rosenblum, Naomi. *A World History of Photography.* New York: Abbeville, 1984.

Rosenblum, Walter and Naomi and Alan Trachtenberg. *America and Lewis Hine.* Millerton, New York: Aperture, 1977.

Rosenfeld, Paul. "American Painting." *The Dial* 71 (December 1921): 649-70.

Rosenzweig, Phyllis D. *The Thomas Eakins Collection of the Hirshhorn Museum and Sculpture Garden.* Washington, D. C.:The Smithsonian Institution Press, 1977.

Rosler, Martha. *3 works/Martha Rosler.* Halifax: Press of the Nova Scotia College of Art and Design, 1981.

Rybczynski, Withold. *Home: A Short History of an Idea.* New York: Viking Penguin, 1986.

Salvioni, Daniela. "Cady Noland: The Homespun Violence of the Hearth." *Flash Art* (October 1989): 129.

Sayler, Oliver. *Revolt in the Arts.* New York: Brentanos, 1930.

Schleier, Merrill. *The Skyscraper in American Art.* Ann Arbor: UMI Research Press, 1986.

Sischy, Ingrid. "Belief." [Dan Weiner] *The New Yorker* 64 (22 May 1989): 79-82.

Spark, Muriel. *Mary Shelley.* New York: New American Library, 1987.

Squiers, Carol. "Barbara Kruger." *Art News* 86 (February 1987): 77-85.

Strandh, Sigvard. *The History of the Machine.* New York: Dorset Press, 1979.

Stillgoe, John. *Metropolitan Corridor: Railroads and the American Scene.* New Haven: Yale University Press, 1983.

Sturman, John. "Medrie Macphee: 49th Parallel." *Art News* 88 (Summer 1989): 174.

Stone, William. "Fall Line." [Artist's Statement] New York: Tom Cugliani Gallery, 1989.

Swenson, Gene R. "What is Pop Art?, Part II: Stephen Durkee, Jasper Johns, James Rosenquist, Tom Wesselman." *Art News* 62 (February 1964): 40-43; 62-67.

Tashjian, Dickran. *Skyscraper Primitives.* Middletown, Connecticut: Wesleyan University Press, 1975.

Tichi, Cecelia. *Shifting Gears: Technology, Literature, Culture in Modernist America.* Chapel Hill, North Carolina: The University of North Carolina Press, 1987.

Trachtenberg, Alan. *Reading American Photographs.* New York: Hill and Wang, 1989.

Tsujimoto, Karen. *Images of America: Precisionist Painting and Modern Photography.* San Francisco: San Francisco Museum of Modern Art, 1982.

Tuchman, Maurice. *A Report on the Art and Technology Program of the Los Angeles County Museum of Art, 1967-1971.* Los Angeles: Los Angeles County Museum of Art, 1971.

Van Wagner, Judy Collischan. *Broken Wings: Donald Lipski at Grumman.* Brookville, New York: Hillwood Art Gallery, Long Island University, 1987.

———. *Walter Murch.* New York: Whitney Museum of American Art at Phillip Morris, 1986.

Vos, George W. [*"Letter to Mr. R. J. Coady."*] *The Soil* 1 (December 1916): 16-18.

Wechsler, Jeffrey. *Abraham Joel Tobias: Sculptural Paintings of the 1930s.* New Brunswick, New Jersey: Jane Voorhees Zimmerli Art Museum, Rutgers, The State University of New Jersey, 1988.

———. *Surrealism and American Art: 1931-1947.* New Brunswick, New Jersey: Rutgers, The State University, 1977.

Wei, Lilly. "The Peripatetic Artist: 14 Statements." [Orshi Drosdik] *Art in America* 77 (July 1989): 136, 155.

Wilke, Hannah, and Thomas H. Kochheiser, ed., with an essay by Joanna Frueh. *Hannah Wilke, a retrospective.* Columbia: University of Missouri Press, 1989.

Wilson, Richard Guy, Diane H. Pilgrim, and Dickran Tashjian. *The Machine Age in America 1918-1941.* Brooklyn, New York: The Brooklyn Museum in association with Harry N. Abrams, 1986.

Wolf, Christa. *Accident.* New York: Farrar, Straus and Giroux, 1989.

Wolff, Janet. *The Social Production of Art.* New York: New York University Press, 1984.

Zabel, Barbara. "The Machine as Metaphor, Model and Microcosm: Technology in American Art, 1915-1930." *Arts* 57 (December 1982): 100-105.

Zaya, Octavio. "Social Security: An Interview with Tishan Hsu." *Balcon* 4 (April 1989): 108-21.

Katonah Museum of Art Board of Trustees

Inaugural Exhibition Patrons Fund

Photographic Credits

Photographs of the works of art in this exhibition have been supplied, in many cases, by the owners or custodians of the works, as cited in the captions. The following list applies to photographs for which an additional acknowledgment is due:

Richard Benson: pl. 27
Geoffrey Clements: pls. 22, 25, 75, 97
Sheldon C. Collins: pl. 16
Joe Coscia, Jr.: pls. 101, 102
Jim Frank: pls. 13, 32, 49, 73, 74, 82, 83, 84, 91
Courtney Frisse: pl. 77
Helga Photo Studio: cover
Donald Lipski: pl. 45
Nathan Rabin: pl. 92
Spectratone: frontispiece, pls. 3, 10, 31, 42
Joseph Szaszfai: pl. 100
Bruce Wall: pl. 93

plate 103
Nina Yankowitz and Barry Holden
Untitled: Painting with Moving Sphere, 1990
fresco, mixed mediums, 2′ x 5′ x 27″
Collection of the artists